The Cellulite Myth

IT'S NOT FAT IT'S FASCIA!

Ashley Black

with

Joanna Hunt

A POST HILL PRESS® BOOK
ISBN: 978-1-68261-288-0
ISBN (eBook): 978-1-68261-289-7

The Cellulite Myth:
It's Not Fat It's Fascia
© 2017 by Ashley Black
All Rights Reserved

Cover Layout: Lunchbox Design, NYC

Creative Director for Cover: Olivia Henry

Interior Design and Layout: Spiro Graphics, Inc.

Post Hill

PRESS

Post Hill Press®
posthillpress.com

Published in the United States of America

Disclaimer

Because we have to say it . . . the information in this book is not intended to treat, diagnose, or prescribe any condition and is for educational purposes. Additionally, this information is meant to supplement, not replace, the advice of a trained health professional. The Publisher and Authors disclaim all warranties, without limitation, and neither the Publisher nor the Authors shall be liable for any damages arising from the use and/or application of the contents of this book, either directly or indirectly. In other words, please use common sense in the way you take care of your body. We are currently researching to scientifically identify key health markers associated with the FasciaBlaster®, therefore all examples in this book are based on anecdotal evidence. Nothing in this book should be construed as medical advice. You should consult your physician prior to undertaking any diet or exercise program, including but not limited to the activities described in this book.

TABLE OF CONTENTS

Ashley's Thanks

First, I want to thank God. There is no doubt that His plan was to lay every stepping-stone in my path to do everything I am doing. My awakening from my near-death experience is the single biggest event in my life, and the divine knowledge I acquired as a result is the reason I am helping so many. My gifts to the world are all His power; I am His messenger. Praise Him.

I want to thank my parents, Bob and Barbara Merrill and Don and Peggy Stansell, who encouraged my business endeavors from the time I was 12 years old and began painting hair bows. They knew I was a crippled child but treated me like an athlete. Most important, they dragged me to church and always allowed me to be me.

I want to thank my brothers and sisters, Reese, Anna, Billy, Mike, and Blair, for challenging me throughout childhood, and for being so different and diverse that we were each able to find our own ways to uniqueness. I also want to thank my entire extended family for their lifelong love and support. A special thanks to the Knight family, Tony, Arien, Micheal, and Saskia, for showing me that life has no boundaries.

Most important, I would like to thank my children, Ryan, who embodies marching to your own fabulous drum, and who inspires me to think globally and without limits; Maddie, who is a ray of sunshine in my life every day and who's leading the next generation of young ladies to a much better place; and Luke, who loves Fasciology and who always brings energy and life to my space. Without my children, I would not have the skills or the love and compassion it takes to "be me" and give the way that I do . . . children teach you that. They have sacrificed their time with me, supported my mission, and were always "good kids." Now they are coming into their own and I can't wait to see the impact they will have on the world. I also consider Layne and Steven to be my own, even though they came into my life later on. Steven jumped in and became the big brother, the "ride or die" and "let's do this" person in my life. Layne, who has been by my side for 15 years, is the most loyal, loving and best "gift" I could ever have. Layne also has performed every facet of the business from day one and now does a stellar job as my mini unicorn, Chief of Staff. I also want to thank my "little sister" Olivia Henry, who is now a part of our Island of Misfit Toys, who contributes to my life and this mission in too many ways to mention; thank you for being my sunshine.

Professionally, I want to thank my very dear friend and co-author Joanna Hunt, for believing in this project for five years and jumping off the ledge with me to make it happen. No one can put my heart and brain on paper the way you do. To my partners, Jeff Pennington, Mike Rome, George Kurisky with MCAP, and LBI Entertainment; specifically Sam Warren and Chuck Pacheco (who stepped up to represent his only unknown client). A special thanks to the work of Jeff Pennington, who came out of a happy retirement to jump on this crazy train. To Shannon Marvin, my book agent with Dupree Miller, who took a chance and pushed to make this happen. To Shelly Reinstein at Autumn Communications and the team at Mills Entertainment. To Anthony Ziccardi and Post Hill Press® for stepping up to publish this manuscript. Scientifically, I want to recognize contributors and my protégés; Bart Jameson, Kathleen Stross, and Dr. Swet Chaudhari MD who jumped on the daily grind to support the science behind this book. Thank you to the team at Grow Online™, specifically Kim Kelley for her unique and innovative approach to changing culture. To my team at Optikal™, specifically Joel Otten, for always taking the risk and pulling off the impossible. To The Woodlands SMTC IT firm for their sleepless nights handling our growth. To Gary Young, for bringing the FasciaBlaster® to fruition and to Gary Solomon for protecting it. To Tori Baxter, who takes care of me so I can do what I need to do. To Enga Sherman, my close friend and assistant, who's been kicking it with me for a decade. There are so many more, and to each and every person who has touched my life, my career, and my soul—I am eternally thankful.

To my FasciaBlaster® sisters, who compel me to keep pushing, and remind me daily of my "why." A special thanks to my Moderators,

Pam, Ella, Janette, Kim, Stacey, and Chelsea, and my entire group of Motivators. To Dea, Johlae, and Sherri for their extra love and support. Without my Blaster Sisters™, this book would never have made it to market, and it's your love and adaptation of this movement that will pay it forward to help millions. The word gratitude is not worthy of the depth of feelings I have towards my **#BlasterSquad™**.

Thank you to two amazing men who have passed on, Sam Aloni and Dan Hoyt. Both amazing men who brought me core business concepts that I use today. Your spirits will live on forever and your impact in my life created a ripple effect. Thank you for being the "drop."

Thank you to Daniel, who from the sidelines inspired me, by just allowing me to be me during the writing of this book. Thank you for sparking my creativity, paddling out, and "my magic bracelets."

SPECIAL ACKNOWLEDGMENTS:

Illustrations: Steven Cherry, Madeleine Ignon, Kim Kelley, and
    Bart Jameson
Cover Layout: Lunchbox Design, NYC
Creative Director for Cover: Olivia Henry
Hair and Makeup: Kelly Thorpe
Editing and Layout: Post Hill Press® & Spiro Graphics, Inc.
Internal Modeling: Olivia Henry
Photography: Debbie Porter Photography

Judy Gass
June 29

"I've never been in a more loving supportive, amazing, sharing group as this one—AND we are a BIG group. I'm so glad to be a part of Blaster Sisters! Blasting fat cells and kicking cellulite's butt. I just want to give every single one of you a great big hug! Love you all. Thank you for the motivation. You guys are the best or as we say in Alabama, y'all are the best."

This book references a number of third-party trademarks. A complete list of these third-party trademarks and their respective owners may be found at the end of this book.

Foreword

For years, as long as this book has been on my "to do" list, I have wondered who I would ask to write my foreword and how I would tee up the information I wanted to share with the world. The foreword, for the most part, is designed to give the author credibility with the reader. I could have chosen an Oscar™ winner. I could have chosen a Hall of Fame athlete. I could have chosen from among a pretty significant pool of billionaires or royals, all of whom are a part of my elite clientele. However, it just never felt right. In fact, it has always bothered me that the rich and famous were the only ones getting "Ashley-ized™."

When I think about my brand and my mission in the world, it's to give the everyday woman the opportunity to *feel* like a celebrity, to have access to the knowledge that the rich and famous have, and to be treated like royalty. In staying true to my brand, instead of featuring someone you are familiar with, I wanted to feature someone you could identify with.

I ran a contest among my private Facebook® group, which is the largest female support group on Facebook® with over 100,000 members, at the time this book was printed. My team and I selected the testimony of Janette Grape. Janette is one example of how so many women have a "crazy train" that runs through their heads about the way they see their bodies, no matter if you're a fitness model, overeater, undereater, California Girl, or Jane Doe in Iowa.

Janette is a physically beautiful woman with a darling and idyllic body shape by any standard. Yet she was imprisoned in her mind because of her twisted body perception and her feeling of being out of control. Her amazing story of self-discovery, empowerment, inspiration, and the drive to pay it forward is the story of every woman. We are all on a journey to tap into our true beauty, inner and outer—beauty that comes from knowledge, empowerment, and ultimately, self-love. Janette now has control over her body and her mind, and she has set up the rest of her life to be the best of her life. Here is her story of transformation, which could very well be yours.

Testimony of Janette Grape

"At the time when I first heard about Ashley Black, my already out-of-control life was spiraling downward. My body image was beyond low. I hated myself, but you would never know it by looking at me. I was in year five of trying to recover from an eating disorder that I had suffered from for over 27 years. As part of my therapy, and because I no longer wanted to live in the body of a 13-year-old, I began lifting weights. Because of the abuse my body endured, which included starvation, cutting, prescription meds, excessively overexercising, and three babies, I was noticeably riddled with 'cellulite.' It became more extreme as I began to put on much-needed weight. Just the thought of what I had done to my body brings me to tears even now as I write this. Looking back, it was never really about being skinny, fat, muscular, or lean; it was about feeling out-of-control with a total lack of education about my body.

"Because of my disorder, I had already gone through losing my job, my house, my marriage, my kids, and at one point I was five seconds from being homeless. The doctors had diagnosed me with sudden cardiac death syndrome, and told me not to workout or even work, because I could die just sitting there. At times I was beyond despair and convinced the world would be better off without me anyway. Can you imagine something as common as cellulite could trigger such a monumental downward spiral?

"Somehow by the grace of God, my husband and I came back together and remarried. I had come a long way, but still my body was a mess, my head was a mess, and truth be told, my life was still a mess. I pressed through and became a little more stable in my health, but then I became out-of-control obsessed with getting rid of cellulite. At times, I would ball up crying because I just couldn't figure out my body, and my husband was at his wits end because he just couldn't figure out me. We were on the verge of a separation once again.

"My misunderstanding about my body, and that cellulite isn't fat, was the root of all evil for me. One day, as I picked apart my every dent, line, and dimple, my husband said to me, "I don't think you understand how much it hurts me—when I look at you I see a beautiful woman and all you do is pick on her. I don't like the way you treat my wife." That hit me hard. It forced me to see that the way I see myself and treat myself affects other people.

"Soon after, I saw an ad for the FasciaBlaster®. I wasn't terribly hopeful but thought, 'Why not?' I've tried everything else.

"It was literally . . . an answer . . . to prayer.

"After I received my FasciaBlaster® and started using it, I had some questions, so I messaged Ashley with my photos, as she encourages. I was blown away when she sent me a voice message herself with her response! I couldn't believe she was so kind and personal. I trusted her right away, which is strange for me, and she began coaching me on my progress.

"As I set out on my blasting journey, my transformation didn't happen overnight. Little changes snuck up on me like winks from Heaven. It all began with self-awareness. As I touched, nurtured, and cared for every part of my body with the FasciaBlaster®, I would cry, because I had been so hard on a body that loved me so much. My body was fighting to keep me alive, and I was fighting it. I began to appreciate how amazing my body really is, and I apologized to it for hating on it. I

> My body was fighting to keep me alive and I was fighting it.

began to see that I didn't have any reason to feel shame. Learning about fascia was an exciting discovery! I didn't know anything about it before or its implications on my body. The FasciaBlaster® gave me what I was lacking—control—and with the knowledge of fascia, I was taking control in a healthy way. At the same time, I noticed that my health was improving because my thoughts about myself were improving. Because my thoughts about myself were improving, my relationship with my husband was improving. My fascia, which had been holding on to so much self-hatred and anger, was smoothing out and releasing locked away emotions. Cellulite was disappearing. I corrected body mechanics, and for the first time, my posture is now as they say, 'on fleek!' My workouts are finally yielding results, because of healthy fascia, muscles, nerve activity, and blood flow. Not only is my body better, my relationships are better, and my life is better.

"The real game-changer was joining the private FasciaBlaster® Facebook® group. Here was this amazing group of women, loving, supporting, encouraging, and accepting one another. Say what?? Vulnerable, strong, courageous women showing off their flaws and feeling safe enough to find support and help. There are no words to describe the profound impact of such a group! You see, my disorder was rooted in severe rejection from women at an early age and repeatedly throughout my life. To find acceptance and love among women of all shapes and sizes was beyond me. I was like, 'Is this even real?' As I would scroll through the pictures I would think, 'These women are so gorgeous and brave, yet they are freaking out about the same body issues that I'm freaking out about.' Not only was I encouraged to know I'm not alone, but anytime I felt badly about myself I would go to the group and encourage someone else in their journey. As I gave love and support, it would always come back to me. Staying in this group is the reason I have progressed to the level I have and my disorder is in the dust. Every woman deserves to feel as connected, supported, loved, and free as the women in this group. It's truly extraordinary and the movement is growing. Like an exotic and rare flower, it just keeps branching out. Ending cellulite and empowering women is undeniable and unstoppable.

"My journey is not over. I thank God for that. I have a ways to go, and Ashley's protocols will always be a part of my life. Her methods are doing for me what over 27 years of searching, being hospitalized, seeing therapists, and taking prescription medications couldn't do. This process is helping me free myself from the prison of self-hatred and a life-long eating disorder. I can't begin to describe how thankful I am for having this information and for being able to care for my body properly for the first time ever in my life. All I can say is thank you. First to God and then to Ashley Black. My life will never be the same because of the in-formation you are about to read in this book, and yours won't either. My story is the extreme, but every person who reads this book will benefit as they experience a shift in the way they approach their body. No matter what the back story, embracing your fascia means embracing freedom— freedom to live, love, and be yourself."

In this photo, Janette and I goof off after an intense FasciaYoga™ session at my home for a FasciaBlaster® retreat.

Introduction

Welcome to the Party!

If you are reading this book, you are in one of two categories of people. Number one: You are a full-on Ashley Black fan and blasting enthusiast; a member of the beloved BlasterSquad™! You likely have gobbled up all the blogs, videos, radio shows, and posts you can possibly find, and you have been waiting for the release of this book like it's Christmas . . . times ten . . . on steroids. You binge scroll through the FasciaBlaster® private page on Facebook® into the wee hours of the morning, and you've seen thousands upon thousands of mind-blowing before and after images of all body parts. All . . . body . . . parts. You may have even been brave enough to post your own before and after images, and felt the massive self-esteem boost when you were subsequently flooded with affirmation and encouragement from women all over the world; women you didn't even know. You've been touched by the love and support in this group, and you've discovered that being vulnerable with other women isn't so scary after all. You may have been part of a CrashAndBlast™. You may have been to a MeetAndTreat™. You have smoother skin, less pain, and you're exercising more effectively. You are changing your body and ultimately your life. You feel empowered in ways you never have before. You no longer punish yourself, because you don't look like the woman in the magazine or "measure up" to society's ideal; you are creating your own ideal. You're falling in love with your body perhaps for the very first time, and that body confidence is seeping over into other areas of your life, too. Your work performance is better, you're sleeping more deeply, your relationships are stronger, and you just feel like a younger, better version of yourself—empowered.

You are a microcosm of a movement of women sweeping the globe—a unique sisterhood, of strong, brave, beautiful women who are tired of being told what they should, or shouldn't, look like and what they can or can't do with their bodies. Women who are tired of competing, comparing, and criticizing and ready to love, share, encourage, and make a difference in their world.

If you have no idea what I'm talking about then you clearly fall into the second category of people. Maybe you found this book in the airport, or a friend gave you a copy. Maybe you saw me on TV somewhere or simply happened upon some click-bait. You were intrigued by the cover and started reading because hey, if there's really a cure for cellulite, you'd be interested. You are a fascia newbie—a B2B™, as in blaster-to-be. Don't worry about what all of that means right now. While the cure for cellulite seems like a handy thing to know, you are about to learn infinitely more about yourself and your body. Infinitely. The information in this book can change your paradigm and impact the way you approach health forever. Now that may seem strange for a book about cellulite to be life-changing, but just stick with me for a minute. I promise to deliver. It's not that we're all so fragile that a little "cottage cheese" on the leg is holding us back from accomplishing our dreams and goals. (And why the heck is a book about cellulite even talking about dreams and goals?) Here's the thing . . . and read this next part carefully: *The key to the cure for cellulite is also the key to a myriad of unanswerable questions about your body, health, and appearance.* In fact, it's a "master key" that's as crucial as knowing that you need water, oxygen, and blood to survive. Once you have this key, you can open as many doors as you'd like. You can change your body in ways you never thought possible. You can correct health and orthopedic issues you were previously told you couldn't. You can obliterate cellulite, pain, restrictions, and more.

In every field of science, there will always be new information. It used to be that these cutting-edge discoveries were first available to society's elite—the rich and famous—and then, over about a 10-year period, it would trickle down to the general population. With today's technology and social media, information is made available globally almost instantly with the click of the "share" button. As we begin this journey together, I want to encourage you to open your mind to the possibility that there is so much more to learn about your body than you've been told before . . . because there is!

OPEN YOUR EYES AND JUMP

In the box office hit *The Matrix*® one of the most iconic scenes is where our hero, Neo, is being invited on a journey that will radically alter and contradict everything he knows about life. He has a choice to make. The character Morpheus offers him two pills, a blue one or a red one. The blue pill will erase the fact that they ever met and any awareness Neo has about what is known as the Matrix®. With the blue pill, Neo can go back to his ordinary life working in an ordinary cubicle and live in the status quo like everyone else. The red pill, however, is another story. The red pill will open his eyes to live fully in reality. It will allow him to see and understand the Matrix®, which is keeping the human race unknowingly in a deep sleep and disillusionment. With the red pill, he will wake up and gain control of his life for the first time ever. Just before Neo chooses one of the pills, Morpheus says:

> *I imagine that you feel a bit like Alice traveling down the rabbit hole . . . You have the look of a man who accepts what he sees because he is about to wake up . . . You are here because you know something. What you know you can't explain, but you can feel it. You don't know what it is, but it's there. It is this feeling that has brought you to me. Do you want to know what it is? It is a world that has been pulled over your eyes to blind you to the truth. You take the blue pill and the story ends. You take the red pill and you stay in wonderland and I show you how deep the rabbit hole goes. All I'm offering is the truth, nothing more.*

In this book, all I'm offering to you is the truth about your body. Nothing more. As you read, you may feel a bit like Alice traveling down the rabbit hole, but hold on because you are about to wake up to some pretty exciting possibilities. You are reading this book because you know something. What you know you can't explain, but you can feel it. You don't know what it is, but it's there. You know there's more out there than what you've been told and now you have a choice to make. Open yourself to the truth and allow me to show you how deep the rabbit hole really goes. You may tumble, reread, question yourself, or feel as shocked as the "BlasterSisters™" who came before you. All I'm asking you to do is trust the process and enjoy the "free fall." This book isn't just about changing your belief about cellulite and fixing it; it's about taking back control. At the surface, it's about taking back control of your body but ultimately it's about taking back control of your life. It's your choice. There's no red pill or blue pill, only the "Black" pill. Go ahead, take it, and jump. Welcome to the party!

1

Leprechauns, Unicorns, Cellulite & Pixie Dust

"It's easier to fool people than to convince them they have been fooled."

−Mark Twain

Cellulite doesn't exist.

There, I said it.

In fact, it's as much of a myth as unicorns and leprechauns. (We'll leave Santa for another book.) The cellulite myth, however, may be the worst of all because this myth has been believed by so many and its implications deeply etched into the hearts and psyche of women.

It's no secret that women are always critiquing their bodies. Two women can have the same size and general body type and both will have a completely different opinion about it. You might hear one woman say, "I love my curves," and another one say, "My butt's too big." One woman says, "I love my arms. I just want more tone," while another says, "My arms are too fat." One woman says, "I love my thick thighs," and another says, "I hate my thunder thighs!" You will always hear a variety of likes, loves, or dislikes about a woman's shape or size, but one thing you will NEVER ever hear is, "I love my cellulite!" No matter what body size, shape, or weight a woman wants to be, no woman EVER wants to have more cellulite. As a "juicy" girl myself, I can appreciate what the Beyonces, J Los, and Kardashians of the world have done, but I don't even think THEY can make cellulite en vogue.

In a seemingly cruel twist of fate, as much as we all hate it, almost all of us have it. In fact, 90 percent of women do. Active women have it, sedentary women have it. Skinny women have it, curvy women have it, young women have it, older women have it, tall women, short women, and even babies can be born with it. You are not alone! But how is this possible when women collectively spend more than a billion dollars a year on "solutions" such as creams, lotions, pills, and procedures? That's billion with a "b." We have talented surgeons, dermatologists, estheticians, alternative care specialists, trainers and nutritionists, supplements, dry brushing, needling, hormones, cupping, essential oils, home remedies, and bloggers galore all working to address the problem. However, there just hasn't been a widely accepted, consistent, affordable, universal solution. (Until now. Keep reading.)

Surprisingly, cellulite has even been labeled a medical condition! Yes, it is actually considered a medically diagnosable condition that almost all women have, and you can't even qualify for medical leave because of it! A med-ic-al-ly di-ag-nos-able CONDITION!!

For years, we've been conditioned to believe that cellulite is some sort of fat issue. So every day, women across are the globe are knocking on wood, wiggling their noses, and slapping some cream on their thighs in hopes that they will one day look good in a bathing suit. (And stuffing themselves into Spanx®.) The problem is that fat is not the problem. Yes, you read that correctly. Fat is not what's making everyone dimply and dented. In fact, there is absolutely no chemical difference between the fat cells where you see cellulite and the fat in any other area of the body. So what does give us bumps, dimples, and cottage cheese? I'm so glad you asked because that's exactly what I'm going to tell you right now! Are you ready? The appearance of cellulite is not about fat, it's about the connective tissue that surrounds the fat called *fascia*. That's why the ultrathin Victoria's Secret® models struggle with cellulite as much as the contestants on *The Biggest Loser®*. I know because I've personally worked on their bodies. Clumps of fat cells aren't dimpled, dented, or lined; HOWEVER, clumps of fat cells bound by tight, sticky, webbed fascia are. If this is the first you've ever heard about fascia, you are not alone. It's one of the most understudied, yet arguably most important systems of the body. It affects virtually everything about your body and for sure, it's what is causing your cellulite!

Now, I'm not the first person to ever make the link. In fact, in a recent Google® search for "cause of cellulite," WebMD® showed up in the #1 placement. Their article about cellulite states in the very first sentence:

> "*Its name makes it sound like a medical condition. But cellulite is nothing more than normal fat beneath the skin. The fat appears bumpy because it pushes against connective tissue, causing the skin above it to pucker.*"

Did you see that? Connective tissue, aka fascia, is what's causing the fat to pucker. That's exactly what I discovered as I worked with my clients. WebMD® is on the right track when they present the notion that connective tissue is linked to cellulite; however, in their article they go on to suggest that in order to reduce cellulite, you have to reduce that fat, which is exactly what most cellulite solutions suggest. However, what if we changed the condition of the fascia instead? What if we simply smoothed out the fascia like one would smooth out a bedsheet? Would the skin on top become smooth, too? Ding! Ding! Ding! Ding! Yes. It. Would. **#DroptheMic**

So why aren't all of these cellulite "experts" telling people to change their fascia? Well, first of all, most people don't even know that they have fascia, let alone how they can change the condition, shape, and appearance of it—but all of that is about to change! Before we go any further into learning about fascia or how to fix your cellulite problem, I'm going to first ask you to *unlearn* a few things. In this book I'm going to expel a lot of myths, and the first myth is the myth that we know all there is to know about cellulite, or the human body, or ANYTHING, for that matter. We don't.

Now, I have a confession to make right at the start: I am a huge research nerd. Most people don't look at a bubbly, gregarious, 5'2", blonde-haired, hazel-eyed, curvy woman and think, "Look at that scientist!" But it is what it is. Having said that, you don't have to be a scientist, like me, to get through this book or learn how to get rid of cellulite. I realize that you may just want the "how to" and not the "why"; however, understanding some of the "why" will enhance your "how to." In other words, I'm going to ask you to put on your big girl pants AND your big girl white lab coat, and learn some of the science. Please. But take a deep breath because if you're not a research nerd like me or if you don't really have an extra brain cell to dedicate to science, let me assure you that this book will make it all super easy to take in and hopefully whet your appetite to learn more about your #**amazingbody**!

On the other hand, if you are a scientist like me, "Nerdy by Nature™" (and comb through obscure studies in foreign countries, spend your vacation looking for aliens, and your idea of a great Friday night party is looking through a telescope), strap in, because you will probably connect the dots early on and be able to see just how deep the rabbit hole goes. No matter where you're coming from, this book has both the "girlfriend talk" and the "science girl talk" so you can jump down the hole as far as you want to go, whether you just want to get rid of cellulite, or explore with me to infinity!

SCIENCE IS ALWAYS CHANGING

Imagine for a moment that you are in a hospital ready for a C-section. You are gowned up, lying on the table when your doctor walks in. He's not gowned up and his hands are filthy with residual blood in the cracks of his fingernails. (Gross, I know. That's the point.) He tells you that he just finished an autopsy on a cadaver who had both leprosy and the plague and he is now ready to begin cutting on you. You have all kinds of alarm bells going off internally as he shouts, "Scalpel!" to the nurse.

"But wait!" you cry. "Aren't you going to even wash your hands?"

If you were a patient in the mid-1800s, the answer would be a resounding, "No." In fact, the doctor would be insulted that you dared suggest that he was dirty. After all, he was part of the medical elite. You see, back then, no one had ever heard of a germ before, even though they existed everywhere, particularly in the cracks of the doctors' fingernails. There were no antibiotics and, clearly, no sanitation standards. In any given hospital, the same doctor who performed an autopsy would walk right into the next room and deliver a baby—Without. Washing. His.

Hands. (Uh! Sounds barbaric and crazy! **#BecauseItWas**) And now you know why we have the health department today.

It wasn't until a Hungarian physician named Ignaz Semmelweis (say that ten times fast) proposed this wild notion of hand washing that the medical community began to entertain the idea. Semmelweis was an out-of-the-box thinker and a pretty sharp guy. He oversaw two birthing clinics, one for the rich with standard protocol as noted previously, the other for the very poor, run by midwives who only delivered babies, and they basically assisted mothers in the streets. Interestingly, the street clinic had far lower mortality rates than the sophisticated clinic run by highly trained medical doctors who performed other surgeries and autopsies, as well as delivering babies. This prompted further investigation by Semmelweis and his team and ultimately, the recommendation that physicians should wash their hands between patients with a simple chlorine solution was born. No pun intended. (Okay! Pun intended.)

Unfortunately, our buddy Semmelweis was shunned by his peers for promoting such nonsense. He was completely ostracized and his life spiraled downward after such rejection. Ultimately, he landed in an asylum, where he was subsequently beaten to death. (How's that for rewarding innovation?)

You see, Semmelweis's cutting edge recommendations conflicted with what was standard protocol at the time. They weren't ready for the shift in thinking that was necessary to solve the problem. The industry professionals were actually offended by his assertion that they were dirty enough to kill someone. Eventually, this idea of hand washing gained widespread acceptance, but not until years after his death, when French microbiologist Louis Pasteur confirmed the germ theory and Joseph Lister began researching and practicing hygienic methods with great success. (Yes, as in "pasteurization" and "Listerine®.")

So what in the world does this little history lesson about hand washing have to do with cellulite? A lot actually. You see, what I'm about to share with you may contradict everything you've believed about your body. It may seem as bizarre as hand washing in a hospital in the 1800s. It will go against much of what you've been told up until this point and will contradict the currently accepted science model in health and beauty. At times, you may think what I'm saying sounds as crazy as Semmelweis, and I'm okay with that really. I'm not here to convince anyone of anything. I'm here to educate the interested. I believe you are interested, which is why you are holding this book! (Or tablet, Kindle®, iPad®, or whatever device you happen to be reading on.) I just want you to understand that science is always changing, so we too must be open to change our view, beliefs, and understanding as new information (or old, unheard of information) becomes available to us.

As I said before, it's hard for people to accept the fact that what we've believed is wrong, or at the very least, limited, even more so in our generation. We have this idea that our technology is now so advanced, that we should know everything about everything; however, that's simply not the case. We are all still learning. Say it out loud, "We don't know everything!"

On top of that, I must say that creating a cultural paradigm shift is not exactly an easy task either. Not everyone wakes up and says, "Today I think I'll change the mindset of the world." But I find myself consumed with this thought every day and years ago actually took intentional action on it. The words of Margaret Mead often dance in my head, especially when I think I'm completely crazy—*"Never doubt that a small group of thoughtful, committed, citizens can change the world. Indeed, it is the only thing that ever has."*

World-changers are driven. Often times, they don't even know why. For me, I was driven first to survive, after being born with a crippling disease that was later compounded by deadly bacteria that nearly took my life. Technically, it did take my life but I fought to get it back. However, if I would have followed the dictates of what the science and medical model of the time had proven, I would be in a wheelchair right now instead of surfing in Costa Rica. (A great place to get some writing done!) The details of my story are at the end of this book, but to me, it's not that important. This book really isn't about me; it's about something bigger than me, and truly, bigger than all of us. Let me explain.

For the majority of my adult life, I've been sought out by some of the most influential people in the world. Before I started my own company and began focusing on educating people, I was practicing Fasciology, which is the science and study of, you guessed it, FASCIA. I was doing fascia therapy on pro athletes, celebrities, billionaires, and even royals. For the longest time, I couldn't even believe that I was the body expert. I had just done for myself and my clients what I thought everyone did, which was to find answers through research and experience. Clearly I had discovered "secrets" about the human body, and I had a burning desire to share them with everyone I could. I was frustrated that I could really only reach a handful of people at a time through my centers. Beyond that, I could really only pour all of my knowledge into a few people who were the closest to me. It seemed so inefficient to get this information out so slowly when it could sincerely help every person.

I had an epiphany years ago when I was working with one of the most influential music producers of our time who has worked with virtually every iconic musician of our time. I was sharing my vision for changing the landscape of healthcare with him and he sat me down and said these words to me, "Ashley, when you accept your place in the universe, you will be a lot happier." You see, up until that point I lived so much of my life literally in "survival mode," first trying to get my body to do what I needed it to do, and then as a single mom trying to run a brick-and-mortar small business, my fascia center. I didn't even realize it, but I was also keeping my knowledge in "survival mode" by sharing it with one person at a time. That moment and those words were life-changing. It was like I had been trying to fill a swimming pool with an eye-dropper and he just backed up the water truck for me. I had to shift my focus from day-to-day survival and accept that what I have learned isn't meant to be shared one person at a time. My life wasn't just about my ability to treat others. I literally had to let go of everything else I was trying to do and focus on getting this knowledge out to the masses in the best, most efficient way possible. I had to accept, embrace, and get excited about my place in the universe. As an inventor, writer, motivator, and game changer, I am there. And I'm here to share with you my purpose.

THE RED BOOK

As I began to meditate on the bigger picture, there were other "marked moments" and larger-than-life encounters that compelled me forward. One of those times was when I was working in Hollywood with a top television producer who had an issue in his body that caused him to literally travel by private plane every week for a year looking for solutions. One day, I was treating him, and as my techniques brought him relief, he told me a very interesting story. You see, he also had a military background with the highest security clearance, so he was privy to the "behind the scenes" in Washington. As I was finishing the session, he got up and pulled out a red book that someone had given him years ago. The book had come out in the '60s and there were only about 10 original copies left in circulation. He handed me a reprint of the book and said, "I'm not sure if you're doing the same thing as the women who wrote this book, but it seems similar." The author of this red book was a Medical Doctor who worked on President John F. Kennedy. She performed a type of myofascial therapy on him, and this book explained her methods. She talked about trigger points along the fascia lines and how to relieve pain. From what I read, she was probably 90 percent accurate according to what we now know today. This woman became the director of the number one medical research facility in the country, the Walter Reed® National Military Medical Center in Bethesda, Maryland. She lived an accomplished life, but what I find so alarming is that here was this Medical Doctor doing what was probably the most alternative medicine of the time and working on one of the most iconic presidents in history and today no one knows who she was. She had about 10 protégés who were studying under her and learning her methods (two less than Jesus) and she wrote this amazing book about what she was doing. YET somewhere along the way, her protégés, protocols, and knowledge fizzled out and her discoveries never achieved mass adaptation.

> "God doesn't call the equipped; He equips the called."

After hearing that story and reading her book, I resolved in my heart that this would not happen to me. I'm not going to spend my life learning and making discoveries about the human body, helping and healing people, and then let it drift off into thin air because I wasn't willing to take the risk, make the sacrifice, or work hard enough to get the message out. I'm not saying that's what happened to her, but I am saying that I am going to do absolutely everything in my power to get this information out.

Her story also made me realize how difficult all of this was going to be. This amazing woman wrote a book, worked on J Freakin' K, and oversaw a research hospital, and even she couldn't get the word out to the masses. It made me stand in awe of all the people in this world who made discoveries in their respective fields, and actually accomplished their dreams and changed the mindset of the masses. It made me stand in awe of my place in my own field of science, which has been growing over time and building into the crescendo that it is today. This information is no longer just for the elite, for presidents, and celebrities; this is knowledge for everyone, and the time to share it is now. I don't want this book to be the "red book" that fell by the wayside—I want this to be the "READ book" that people talk about for ages to come!

They say timing is everything and truly it is. It's one thing to have knowledge about a subject and an uncommon ability in an area—a gift, if you will. It takes more than being a scientist and an inventor to get a message to the marketplace. It takes someone who can also talk to the marketplace. It takes funding, it takes a team, it takes the right exposure, the right partnerships, the right PR, and it takes the right moment in history. You might say it requires a little pixie dust, but what I believe is that it is a Divine ordering and everyone who is a part of this movement feels it. I love the phrase, "God doesn't call the equipped; He equips the called." I believe wholeheartedly that sharing this information with you is my calling. God threw me this ball that I am running with, and now I'm throwing it to you. I'm accepting my place in this world, and I'm inviting you to accept your place as well. It is a place of empowerment and freedom. A place where you know yourself better than anyone else. A place where you make the choices for your life that lead you down the path of joy and fulfillment.

THE WAY OF TRAILBLAZERS

When I think about the great pioneers of the past, I imagine what it was like for them and I find inspiration in their personal stories. Men like John D. Rockefeller. Here he was, a totally uneducated man (unless you count his 10 weeks of bookkeeping when he was 16), who became the world's richest man, controlling 90 percent of all oil in the United States at one point—and the first American worth more than a billion dollars.

Rockefeller faced a lot of hardship in his life. He started his career struggling in the kerosene business. This was pre-electricity and his business was headed nowhere. He also narrowly escaped death when he missed getting on a train that derailed in what was one of the most horrific train accidents of the time. Because of this event, he turned to his faith and became a devout Christian. He then partnered in business with Cornelius Vanderbilt, who owned the railroads. Rockefeller had all this runoff from making kerosene and he needed a way to either use it or dispose of it, so he hired 12 scientists to study it and find out what to do with it. The byproduct was gasoline, which of course changed the world and the way we do everything.

Then there's Walt Disney®, the cartoonist who built an empire now worth over $100 billion, and winner of 22 Academy Awards®. He was a high school dropout who was fired from a job early on for not being creative enough! (**#SayWhat**) People tried to push him down and even stole his ideas, but he just kept going. He was rejected by banking institutions, but he just kept going. He had laser focus and kept his hand in every aspect of his business as it grew. He had a spirit of excellence and a drive that others would call perfectionism. It was revealed in his acute attention to detail that lives on in his company today. No one would argue about the role of Walt Disney® in the fabric of our global culture.

Another inspiring story is the formation of Cirque du Soleil®. It's not that we haven't seen tumblers in a circus, or heard great music, or experienced amazing light shows, and fabulous artistry. However, when Cirque du Soleil® brings together these elements, something very special happens that engulfs the senses and evokes emotion.

> *People might think that we set out to reinvent the circus, and then just did it. But things did not happen that way. We were a bunch of crazy people who wanted to do things, and little by little we came to a vision of what the modern circus could be.*

—René Dupéré, music composer for 10
Cirque du Soleil® shows

Not one of these trailblazers were educated in their respective fields because before them, their fields simply didn't exist. I identify with Rockefeller's struggle to make something with a substance that has been previously ignored. Fascia, for the most part, has been largely ignored and discarded. I can relate to the notion of escaping death through Divine intervention. I respect his unwavering faith and drive to find answers. I honor his desire to create and manage a new industry/genre.

I identify with Walt Disney® when everyone thought his ideas were too big, or too far outside the scope of comprehension. Like him, I've faced the critics and those who want to pirate off the success of others. He always stayed true to his vision and never took "no" for an answer. This drive resonates deeply inside of me. I also believe that if I take the body solutions I have, I can essentially build my own healthcare "Disney World®" where people will come and enjoy it as the happiest place on Earth! (Don't inbox me or ask me when this will happen!)

I can identify with the creators of Cirque du Soleil® because they took artistic elements from across the centuries and connected the dots to create the shows we see today. This is similar to the way I connect the dots with the fascia pioneers before me, and I want to take a moment to give reverence to their work: da Vinci, Rolf, Myers, Schleip, Travel and Simons, and all the others who have served up pieces of the fascia pie. You could say I've revised their recipe and added my own "ice cream" to make it that much better, clearer, and tastier to gobble up! You could also say I am the fascia symphony conductor who is bringing it all together in perfect harmony for the big show. The stage has been set for a long time, and I'm inviting you to come take your front row seat!

No, I didn't discover fascia, but I'm sincerely thankful to be in this moment of history, building on the science and research that created a foundation for me to gain control of my own life. I'm both honored and overwhelmed to be the trailblazer to give fascia a face and a voice. Beyond that, I'm sincerely honored that you're here to read and observe.

Yes, the time is right. The knowledge is in place. This book is about every second of my life making a difference in someone else's life—yours. To get really deep and honest with you (warning: I'm generally really deep and honest), I did not almost die in a hospital bed to come out and do nothing. I am not going to the grave with my secrets.

So yes, in this book you will learn how to get rid of cellulite, but in learning how to get rid of cellulite, you will learn so much more. I'm not here to end all surgeries, or tell people what prescriptions they can or can't take, or say people should fire their trainers, or stop their current programs, or even possibly change their diets. I'm simply asking you to open your mind to the idea that what you've been taught isn't the end of the story, it's only the beginning. We are on this journey of life together, and I'm here to empower YOU to make INFORMED choices.

Every field of science is constantly evolving, and what you are about to read is on the very cutting edge of body performance, and health, and beauty breakthroughs. When you fully grasp the concept of Fasciology and how it can vastly improve the quality of your life, you will wonder why it's not being talked about by every type of therapist, nutritionist, medical professional, trainer, body worker, or other gurus on the planet! Who knows, you might even join the ranks of the trailblazers in your own way too; and at the very least, join the movement and inspire someone else. This is my dream. Thank you for being a part of it.

Tiffany Wilkerson
July 8

I want to tell you a story about a woman who found her life again at 40.

Most women will be the first to tell you that at some point in their life, they look into the mirror and barely recognizing the reflection/stranger in front of them. There is a questioning that arises that asks, "how did this happen?" For me . . . that was my turning point.

Now I've been through the years of having a youthful body growing up, played sports . . . . Then enters the age that I welcomed marriage . . . then 2 difficult pregnancies that involved surgeries . . . then, the post pregnancy era. Now don't think for one second you ever regret the sacrifices and somewhat disfiguration of giving birth and creating a beautiful family, but there is a "physical aftermath" that you are faced with. Once I saw "that," I knew I had to get busy. I began to thrown myself into gyms for hours a day, took on workout routines at home, and more. If you name it, I've probably tried it. But none ever gave the true satisfaction of what I was looking for.

Then one day . . . Ashley Black pops up in my Facebook® stream. Although I didn't understand this fascia blaster tool I was looking out . . . there was a strong sense telling me to give it a try . . . .so I did. And to test it even more, I ceased my workouts, Keith my same menu . . . .all for the chance to see what this thing could do in its own . . . .no distractions. And you know what . . . .? My body transformed more in 2 weeks than the years I was in the gym. This cellulite that I labeled as a "permanent tattoo of life and age," diminished. But the even greater find more than for vanity reasons, I found an ease of pain for an 18-year-old back injury I had experienced from 2 car accidents in my younger years. Can you imagine not waking up in pain after 18 years of trying to "normalize" discomfort? For the first time . . . I woke up smiling after a few weeks using the blaster. No more not wanting to get out of bed on some days . . . no more Tylenol . . . nothing but smiles and I'm never turning back.

Ashley Black has changed my life by giving me a new one and I'm most thankful for it.

If I had the chance to talk to a million women sitting on the fence about this tool, or even just one woman . . . .I'd tell them to do it. Every myth you've ever been told about "it's just in your genes so deal with it", and age makes the body change . . . you'll no longer believe it.

You CAN have a life . . . at 40, 50, 60, post children . . . . .whenever . . .

You CAN live and Ashley Black shows you how to unlock that access into all you ever hoped for or ever lost.

That moment that changed my life was when I found Ashley Black.

✛   🖼   ✳✳✳✳✳

2

Fasciology 101

"Healing is a matter of time,
but it is sometimes also a matter
of opportunity."

—Hippocrates

Have you ever thought about why your liver doesn't crush your stomach? Or why you don't feel your organs knocking around when you jump and run? Or how your 25 feet of intestines don't free float like gummy worms flopping around in your belly? Have you ever wondered why your heart stays in exactly the same place? Or how muscles stay attached to your bones and don't slide down like stretched out socks? Or how your nerves and blood vessels stay in the same place? Probably not, and why would you? Amazingly, in an age where we've put a man on the moon and can open a door with our fingerprint, the vast majority of people have very little understanding about the inner workings of their bodies, especially when it comes to fascia.

You might have heard of connective tissue, but do you really know what that is? What if I told you that many physical pains could quickly and easily be relieved, the aging process slowed down, and yes, even cellulite disappear? Surprisingly, many people have never even heard of fascia, which is technically connective tissue; however, it's so much more. Fascia is a system completely of its own. Fasciology is the science that studies this system.

Okay, I just want to pause here for a minute and ask for your complete, undivided attention. Read my lips: THIS IS THE MOST IMPORTANT SECTION OF THE BOOK! Please do not skim it! Do not gloss over it. Read it carefully, chew it, digest it, take notes, underline, highlight, and reread it if you have to. (Yes, I realize you can't technically read my lips.) But please, please make sure that you are alert and undisturbed while you read the rest of this chapter. I'll even wait for you if you need to go grab a coffee or find a quiet place to focus. This is super important so do what you gotta do to take it all in. Understanding fascia is as important to your health as understanding that you need water to survive!

THE FASCIA SYSTEM

Fascia is a highly sophisticated system of cells that is spread throughout your entire body. It also interplays with every other system of the body—circulatory system, cardiovascular system, muscular system, nervous system, and so forth. It's quite complicated, actually, but I'm going to try to simplify it as much as humanly possible. Imagine if you had not ever seen a computer before and I was going to show you how one worked. Think of this section as a high level introduction to the computer. In just a bit, we'll throw it on the table, pop open the back, and I'll show you the motherboard and hard drive, then explain what you can do with it.

Okay, back to fascia.

Fascia tissue is a sticky, web-like substance that attaches underneath the skin and is literally everywhere in the body. Everywhere. It both separates and connects the muscles and the organs, bones, veins, and arteries that run through it. It both wraps around and penetrates your brain. It runs from head to toe, fingertip to fingertip, and surface to deep. It has qualities like plastic wrap and can be likened to a giant spider web. Some have described the way it runs throughout your body like a system of highways on a U.S. map, with roads and complicated interchanges. It's not just on top of everything; it's under, in between, and throughout everything, which will make more sense as we go along.

If we were to extract the fascia system from the body fully intact, it would look just like a 3D cotton candy mummy. It exists as one single unit without interruption, penetrating, connecting, and separating every single part of the body. If you pull back the skin, the very first thing you would find attached to the skin is a layer of fascia. You would be looking at one of the four types of fascia, which we will get into in the next section.

Here's a visual to help you grasp the concept, but it's still not the complete picture, so stick with me. Back in the old days, before we had every single convenience food known to man prepared for us, back before grab-and-go Jell-O® cups, people actually made Jell-O® at home from a powdered mix and some hot water. We called this cooking. (Yes, please note a small bit of sarcasm there.) Anyway, to make Jell-O® the old-fashioned way, you boil some water, pour in the powdered mix, stir until dissolved, then transfer it to a mold and refrigerate. As the liquid cools, it solidifies into the see-it-jiggle-watch-it-wiggle tasty, fun treat enjoyed by kids of all ages for decades. Now, imagine that you have boiled the water and stirred in the powder, but before it solidifies, you drop a small loofah shower scrubber into the pan. What would happen? The Jell-O® mixture would run *throughout* the loofah because it's porous. The Jell-O® mixture would also *surround* the loofah. So you have this pot of liquid Jell-O® with a loofah sitting in the middle. Then it cools and solidifies. The loofah is now held in place because it is both surrounded by and penetrated by the firm, yet jiggly, Jell-O®. The Jell-O® is inside the loofah, outside the loofah, and connected as a single unit. This is like fascia inside your body that it penetrates and surrounds the internal structures; however, keep in mind that fascia isn't completely solid. It's more webbed like cotton candy. Again, this is a great oversimplification of the fascia system, but hopefully it gives you an idea of how fascia penetrates and connects everything throughout the body.

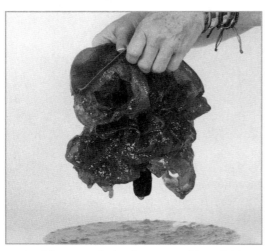

Loofah surrounded and penetrated by Jell-O®
the way fascia surrounds and penetrates internal
structures of the body.

If you search the internet for information on fascia, you're going to find a myriad of opinions about the way the fascia system works, some good and some bad. Fascia research isn't well funded so there are pockets of curious scientists out there studying it, and bloggers are starting to write about it. Even a few major magazines are touching on it. Because the fascia research has come from lots of smaller groups, there hasn't been a cohesive agreement about how the science should be presented or studied further. It's like if you hold a dollar bill up in between two people and ask, "What do you see?" Well, one person would say, "I see George Washington." The other would say, "You're crazy, that's a pyramid!" They are both right, but they are looking at it from different perspectives. And then, of course, there are just some people out there who have no idea what they are talking about. But the general public hasn't been educated enough to know who those people are. As I studied, researched, and looked at the intricacies of fascial dissection, it was very clear to me that it was UNCLEAR how fascia has been presented to the mass market, and it was time for this book. For example, the science is very solid with the dissection work of Tom Myers, who calls Structural fascia "trains and stations," where I just call it Structural fascia. For this book and for my curriculum for Fasciology, I pulled from the legitimate, established science and organized it in a way that I believe the average "non-science" person can understand. Now that I have the opportunity to write this book, I want to deliver the information to the masses and clear up the confusion! There are many fascia "experts" out there so just know if you get Google®-happy, we might all be saying essentially the same thing but using different terminology. Fascia "language" has yet to be established, so I hope this book will bring the fuzziness into focus.

THE FOUR TYPES OF FASCIA

Now it's time to look at the inside of our "computer" and see what it's made up of.

The fascia system is comprised of four types of fascia that are connected as one, however they are found in different parts of the body with slightly different compositions. The reason I want you to have a basic understanding about the four types of fascia is because each type of fascia affects the appearance of your cellulite differently. With this understanding, you can analyze your own cellulite and be able to address it in a more effective way. You will know why you have a dent and be able to determine where it's coming from; you will understand chunks under the skin, or ripples, and why you might bruise or temporarily swell. So while you may not yet be as excited about this as I am, you will be when you have those super smooth legs you've always wanted!

Without further ado, the four types of fascia and their scientific names are:

1. **Structural (Ace® Bandage)**
2. **Interstructural (Cotton Candy)**
3. **Visceral (Goopy)**
4. **Spinal (Straw)**

Okay, so maybe the words in parenthesis aren't technically scientific names, but they are names to help give you a visual so you can remember what I'm talking about. Besides, all scientific names are made up by someone—so we made up scientific names and little terms to make things easier for you to learn instead of making them more complicated. Ace® Bandage, Cotton Candy, Goopy, and Straw Fascia may not sound terribly official, but I'm not trying to be official; I'm trying to help you understand your body. Hopefully these terms will stick in your brain like the fascia it's housed in! (Pun intended.)

TYPE ONE: STRUCTURAL FASCIA (ACE® BANDAGE)

The first type of fascia we are going to explore is Structural fascia and it's the fascia that you see directly under the skin. It doesn't matter if you dissect the head, cheek, back, arms, or feet, this layer of fascia will always be the first thing you see. But it's not just under the skin; it runs in long strips all throughout the body. Think of an Ace® Bandage. It's a long strip of material that's woven and stretchy. In its natural state in the average person, it's as stretchy and as flexible as a thick rubber band yet, if you run your fingers over it, it feels silky and sticky, like plastic wrap. Pictures of these lines of fascia are found on page 118 under "The Position Test" if you want to flip ahead and take a look. (But come back and read the rest of this chapter!)

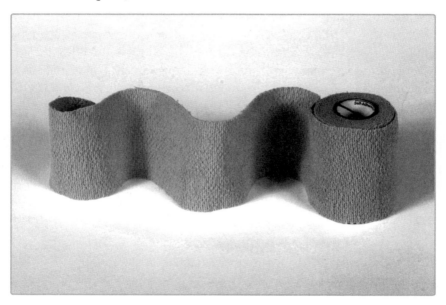

Ace® Bandage depicting Structural fascia

So imagine an Ace® Bandage running all throughout your body. It doesn't run in a straight line either; it runs more like highways on a map with sophisticated interchanges. The important thing to understand is that these lines of fascia are not necessarily on top of muscles or surrounding them. It's more complicated than that. Lines run at different layers or depths. They can run just below the surface of the skin, then travel down inside the body to a deep place, and ultimately resurface at another part of the body. The lines are not organized like layers of cake and icing; it's more like a plate of spaghetti. Another way some people describe Structural fascia is by comparing it to the membrane on an orange. When you peel an orange, this would be the white stuff just underneath the orange's surface separating and connecting the slices. Starting to get the Structural fascia picture?

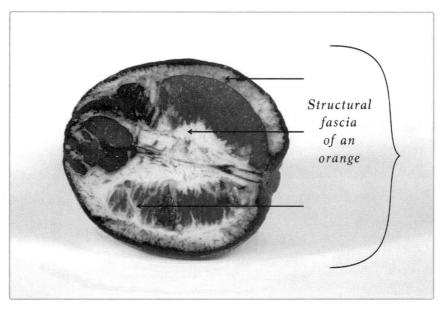

Structural fascia of an orange

This layer of fascia is also visible on a piece of chicken, particularly fried chicken. When you lift the skin, it's the white membrane you see, and sometimes you can see fat mangled into it, which is what gives it the appearance of dimples and wrinkles in the skin. When we get into the types of cellulite, this is the surface layer that causes Hail Damage (see page 98). The Structural fascia can be totally smooth like plastic wrap or totally messed up like a crinkled sheet of paper. Again, this is what you will see in your skin, but it's caused by your fascia.

In anatomy, this layer is referred to as a sheath. I've also heard it called a membrane, and there's another school of thought that calls it the outer bag or surface fascia. (See why people are so confused about fascia?) You just need to know that it is Structural fascia, aka Ace® Bandage fascia—a single layer, in strips, that covers many surfaces inside your body.

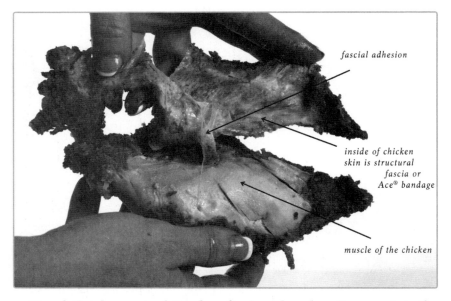

fascial adhesion

inside of chicken skin is structural fascia or Ace® bandage

muscle of the chicken

Now, let's take our Ace® Bandage fascia and see how it connects to the next type of fascia.

TYPE TWO: INTERSTRUCTURAL FASCIA
(COTTON CANDY)

Okay, so this may sound like a preschool art project, but who doesn't love preschool art? (Don't throw this one away.) Imagine taking our Ace® Bandage Structural fascia and gluing cotton candy all over it in every direction. Then imagine THAT whole crazy mess is all over the place inside your body. The cotton candy is, of course, the Interstructural fascia. There is a popular YouTube® video that calls it the "fuzz." I call it Interstructural because, well, it runs throughout the structures of the body—that's why it's "Inter" structural. (Genius, right?) This small, sticky, web-like fascia tissue weaves through virtually every structure, muscle, organ, and so forth, and attaches to the bone. If you cut into a piece of steak and look very carefully, you would see a thin webbing of Interstructural fascia throughout the meat. It's about the size of strands of hair but with some stretch and stickiness to it. The medical community sometimes calls this "fascial bands" but don't confuse it with our Ace® Bandage Structural fascia that runs in bands.

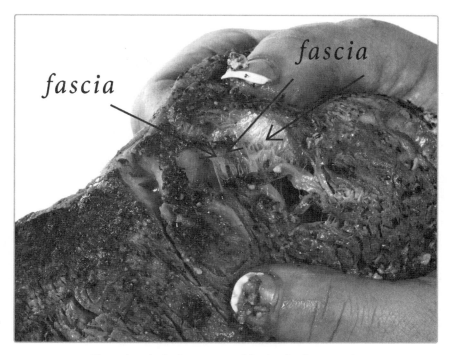

Shown here is the Interstructural fascia of a dinner steak.

In a little while, we will be discussing what happens when the fascia becomes unhealthy, but I want to touch on the concept briefly here because I think it will help you understand the relationship between the types of fascia a little better.

When fascia is in an unhealthy state, it sticks together or tightens around whatever it surrounds and penetrates like two pieces of duct tape stuck together. You might have heard this referred to as a knot in a muscle, but muscle fibers don't tie in a knot; however, fascia does! Fascia can tighten a muscle and basically "glue" the structure into what feels like a knot, but it's really balled up fascia tissue. We call this a fascial adhesion.

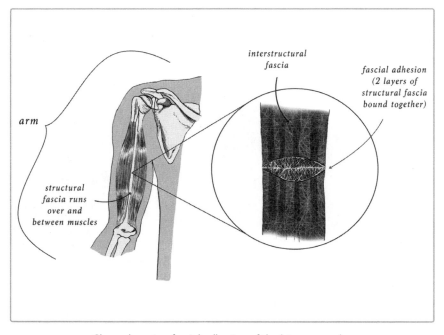

Shown here is a fascial adhesion of the biceps muscle.

The fascia will clamp down around muscles, but when you feel the "knot" it's not a muscle problem, it's a fascia problem. When our Cotton Candy fascia has places that are unhealthy, it pulls down the top layer of Ace® Bandage Structural fascia, creating little dimples and puckers in the skin that it's attached to, much like the tufting technique in furniture upholstery. So imagine that you have the Ace® Bandage laid out with the Cotton Candy hanging from it and you pull down on the Cotton Candy, which pulls down on the Ace® Bandage. Voilà! Cellulite dents and divots are formed. Then you have fat pushing through these dents, which pronounces the look. The Ace® Bandage forms the "valleys" and fat forms the "mountains." Yes, you can put the book down for a second and check your legs. Now you know what's happening inside the deeper layers of fascia!

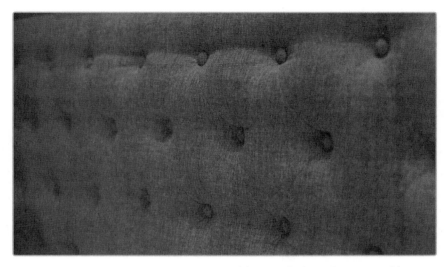

This photo represents how the Interstructural fascia pulls down the Structural fascia

Okay, seestahs, now let's take our art project, the Ace® Bandage with the Cotton Candy hanging from it, and let's run it through a big pile of slimy goop. Doesn't that sound fun? On to our next type of fascia.

TYPE THREE: VISCERAL FASCIA (GOOPY)

Have you ever watched one of those sci-fi alien movies where the creature pops out of someone's abdomen wrapped in a gooey, stringy mess? Introducing the inspiration behind the goop—the Visceral fascia! Visceral fascia—which we will appropriately nickname "Goopy" fascia—is somewhat like teeny-tiny strings of extremely sticky Jell-O® that's housed in the abdominal cavity. Now, the abdominal cavity is hardly an empty place. Remember our Jell-O® and sponge analogy? The same thing is going on here with the Visceral/Goopy fascia and the internal organs, except this type of fascia is a totally different texture from the fascia in other regions of the body, because those vital organs are, well, vital. Fascia is a protective system of the body so there's more of it where more protection is required. We'll talk more about how fascia "protects" in just a bit but for now, let's focus on the Visceral fascia.

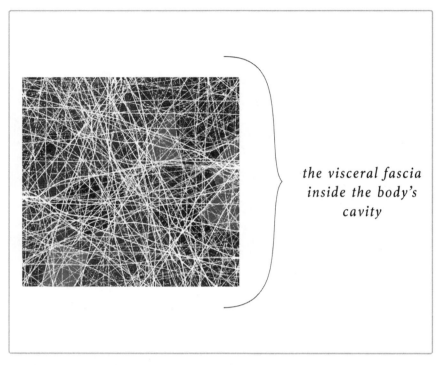

the visceral fascia inside the body's cavity

This illustration is not a "literal" image of Visceral fascia.

If you were to cut open the body cavity and stick your hand into the Visceral fascia, you would find that it feels slimy and wet. When you pull your hand back out, it would be covered with Goopy fascia dripping like strings of melted mozzarella cheese, but still slimy and wet. Anyone who has gutted a turkey on Thanksgiving has had a close encounter of the Visceral fascia kind!

Unhealthy Visceral fascia can greatly affect a woman's ability to achieve a small waistline or get her waistline back after having a baby. Have you ever heard a woman talk about how she had nice flat abs, no stretch marks, no dimples or cellulite on her stomach before having a baby, and then after the baby, her body was just never the same? Here's why: when a woman gets pregnant, her body stretches to accommodate the baby. No surprise there. But what happens *every* time is that the Structural fascia that runs through the abdominal cavity is literally torn and *sometimes* the Visceral and Interstructural fascia are also torn. To say at the very least that the fascia is totally disrupted during pregnancy, or weight gain or loss, is an understatement. After the baby is born and the fascia tries to seal back up, it doesn't necessarily go back together nice and neat the way it was before. This, along with stretched skin, is why women sometimes have droopy, saggy, dimpled, soft bellies after having a baby. Not only that, when fascia is disrupted like that, blood flow and nerves are also disrupted. Muscle access or the ability to contract muscles is impeded, which is why women have trouble firing their inner core muscles and pelvic floor like they used to be able to do.

The Visceral fascia can also present as being overly tight. Men and women experience this in their abdominals. In men it appears as a "beer belly" and women usually look like the picture below. That is what we call Beyond Bound™ Visceral fascia (see page 44), the worst state your fascia can be in. You'll understand more about the health implications associated with unhealthy or disrupted fascia in the next chapter. For now, just let it sink in that fascia is a vitally important component to the overall health picture.

As far as our Goopy, Visceral fascia goes, remember we have two other types of fascia running through it, along with all the vital organs, muscles, nerves, arteries, and so on. To say the abdominal region is a complex part of the body is the understatement of the year! Picture highways and pulleys and levers, all connected in one continuous piece of webbing . . . are you starting to get the picture?

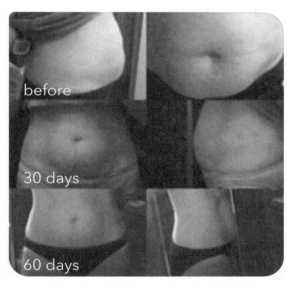

60-day results of restoring the Visceral fascia.

To understand cellulite, which is what you see on the outside of your body, you have to understand a little about what is happening on the inside of the body. Hopefully you're starting to understand how the mess inside makes a mess on the outside. (Take heart, I'll help you clean up the mess!)

Now, let's take a look at the fourth type of fascia.

TYPE FOUR: SPINAL FASCIA (STRAW)

spine of turkey

This image shows the Spinal Straw fascia of a turkey spine.

When I was searching for images of the Spinal Straw fascia, it was astounding to me how far behind we are in our anatomy lessons. If you dissect a human body, you will see that the entire Spinal Straw is encased in a very thick, straw-shaped tube of fascia. This is a unique place where the fascia surrounds something in this manner. This straw is in strips or sheets like the Structural fascia, and there is Cotton Candy fascia and Goopy fascia hanging from it. It's plain as day that this straw of tissue is fascia tissue and part of the fascia system, connected to every other type of fascia in the body. Just Google® "spinal dissection" if you have the stomach for it. Now, I don't normally get as passionate as I am about this topic right now, (okay, maybe I do . . .) but do you know what they call this in traditional anatomy textbooks? They call this "arachnoid mater." ARACHNOID MATER. Which loosely means "spider web-like material." The Spinal Straw fascia in traditional anatomy is broken down into dura mater, arachnoid mater, and pia mater. Dura mater is defined as the thickest outermost membrane that surrounds the brain and spinal cord. Arachnoid mater is the layer in the middle, and pia mater is the thinnest and innermost layer. So basically, what they are trying to say is that there is a layer of fascia attached to a layer of fascia attached to a layer of fascia, structured just like fascia everywhere else in the body. In the anatomy curriculum, this "spider web-like material" is only associated with the brain and the Spinal Straw, even though it connects to all of the other types of fascia throughout the body. These layers of fascia that surround the spinal column can tighten just like fascia anywhere else. (**#SpoilerAlert** . . . you'll find out what that means in the next chapter!) When are we going to get with it and realize it's all part of the same system—the fascia system? Let's call it what it actually is—fascia! (Arachnoid Mater . . . Psh!)

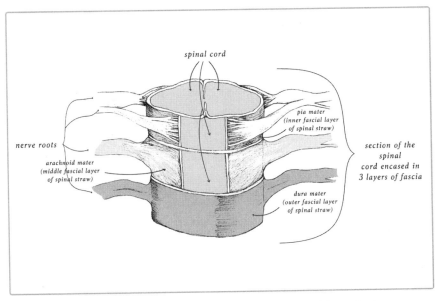

This image shows the 3 layers that make up the Spinal Straw fascia.

Okay, so if we go back to preschool (which is about how far behind the textbooks are on this research), and we take our art project with the Ace® Bandage and the Cotton Candy hanging off it that's been run through a pile of goop, and *then* we attach it to a great big giant straw, we will have our very basic model of the fascia system. When we talk about a narrowing of the Spinal Straw, this is nothing more than a tightening of the Spinal Straw fascia clamping down on the vertebra, which may very well be related to tightness in another part of the fascia. Loose fascia everywhere else supports loose, free-moving Spinal Straw fascia. I will say it again, it's all connected! When you have all four types of fascia in optimal condition it solves *soooooo* many medical, health, and beauty issues. A+ for everyone—let's have a snack and a nap! Just kidding.

PUTTING IT ALL TOGETHER

For now, really take hold of this concept, because all of the fascia is connected, tightness in one area can pull and restrict and change the appearance of another area of the body. So your cellulite in your thigh may or may not be because of tightness in your thigh. It may be because of tightness in your calves, or back, or neck.

Think of a bed sheet laid out on a bed. If you pull on the bottom of the sheet, it's going to affect the placement of the top edge of the sheet. Likewise, your cellulite and your back pain can be related.

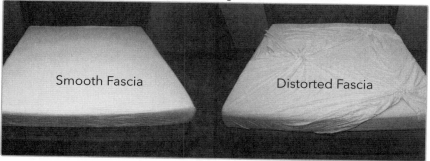

This image of 2 bed sheets represents, on a global level, how distortions in the fascia in one part of the body affect the whole.

Now, before anyone starts inboxing me about any medical conditions, I'm not claiming to have all the answers. The human body is very complex, but we are hopeful about the research connecting the manual release of fascia to the cure for many issues. In fact, my own research is in full swing and I'm dying to get the data! In addition to helping you get rid of your cellulite, my goal with this book and future book projects is to spawn research and awareness about the importance of fascia to the overall picture of health, beauty, and wellness. We've come a long way in our understanding, that's for sure. The question is, are we even looking at all the possibilities with fascia? The same is true for cellulite. We've all been boxed in to believing it's a genetic issue, or a problem that can't be solved, but maybe we aren't asking the right questions and so we aren't getting the right answers or the right results. My mission is to start the conversation and at the very least present the fascial component to all of these conditions, problems, and "unknown" issues.

As you may be starting to realize, fascia has a brush point with literally every other system of the body. It is as significant or *more* significant than any other system.

Fascia, in my not-so-humble opinion, is the most profound system of the body! Science shows that the very first cells of life are fascia cells. What is medically called a zygote is a cluster of living fascia cells, and we are created that way for a reason. We are literally born and formed and develop into our fascia. It's like our root system; everything we are comes from fascia. Consider Psalm 139:9, which says, "You knit me together in my mother's womb." Fascia is what knits us together in the womb. I find all of this so mind-blowing that it's unbelievable that, by and large, we still aren't looking at this system! Fascia cells are our God-given first cells for a reason! There's much more to be said, but it's really not for this book; however, I want to place the magnitude of the impact of fascia on the human body into your lap. **#ThinkAboutIt**

Kiera Leatherman
February 28

Growing up I had chronic fatigue, bulimia and then since my epidural 14 years ago chronic back pain that got more and more intense. I am now a mother of 6. My weight has gone up and down and even at lower weights I couldn't understand why I still had cellulite. I just ached to feel comfortable in my own skin . . . And to love myself.

In May I started Ashley's protocols. I was determined to get to the root of the issues of my health. I helped heal my children from several health issues and it was time I devoted healing to me. I had been trying far too long and needed something more especially for my back and cellulite. Every night I did a section or more. I was devoted and excited about what was happening. I began to have more energy again and was able to begin some buti yoga last week, have completely stopped emotional eating and instead I am nourishing myself in every way! I am also feeling so alive and confident to continue with my business plans!

I am finally falling in love with myself and it's not only because of toning up but also learning that my past wasn't my fault and I'm worth taking care of me and nourishing my body and soul. Tomorrow my youngest is turning one year old and she was my csection baby. I've blamed myself so much. I really felt my body was broken. I no longer feel that way. I feel stronger than ever and healthy. Maybe my csection actually taught me to take care of myself, no one else went through that — not my children, not my husband. I am the one who had to heal from it and I deserve to do whatever it takes to do that. Now I can right at home. It is so empowering to be able to take care of myself. Thank you so much Ashley. I can't thank you enough for the relief this is bringing me in so many ways.

3
Healthy Fascia vs. Unhealthy Fascia

*"It is more important to know
what sort of person has a
disease than to know what sort
of disease a person has."*

−Hippocrates

Okay, class, great job on Fasciology 101! If your head is still spinning, keep in mind that you officially know more about the human body than 90 percent of your friends. Now, what you thought was just a preschool art project is morphing into the secret key to health, beauty, wellness, anti-aging, and vitality! Here's the part of the book where all the dots start connecting and forming a clear picture. Yes, here is your answer to your cellulite questions, but strap in because the rabbit hole is about to take you to Wonderland!

As you might have guessed from the chapter title, now we're going to explore what it means to have healthy fascia vs. unhealthy fascia. By the end of this chapter, you'll understand in detail how the condition of the fascia impacts not just your cellulite, but your overall health!

The fascia system is comprised of cells just like everything else in your body is comprised of cells. Regarding any system of the body: healthy cells equal healthy functioning. When fascia is in a healthy state, it's soft, pliable, supple, and hydrated, and it supports the healthy functioning of the internal structures it surrounds and penetrates. When fascia is unhealthy, it tightens and sticks together, clamping down on everything it surrounds and penetrates (which is everything) and a whole host of problems ensue! It's not an "all or nothing" situation though. You can have healthy fascia in some areas and unhealthy fascia in other areas.

> When fascia is unhealthy, it forms what we call restrictions, adhesions, and distortions or "R.A.D." for short.

When fascia is unhealthy, it forms what we call restrictions, adhesions, and distortions or "R.A.D." for short. (But in this case, it's not rad to have R.A.D.) A restriction is a place in the body where the fascia is tight and restricting something, like blood flow, nerve activity, joint mobility, muscle access or even organ function. An adhesion is a place where the fascia is stuck together and balled up like duct tape, and is commonly referred to as a "knot." (It's not a knot!) Adhesions can cause restrictions. A distortion in the fascia is where the tissue is pulled out of shape or the body's structure is distorted, such as with the appearance of cellulite or when a person has a knee that is turned in. The tight, unhealthy fascia is pulling and distorting the anatomy. There are cases where the body is riddled with R.A.D. in a more systemic way, but most people have varying degrees of R.A.D. in different areas of the body. In the next few chapters, we'll look at how to analyze your fascia and bring it to a healthier state. When fascia is brought from an unhealthy state into a healthy state, we call this "restoring the fascia." Remember that term. The images on the next page show the difference between healthy and unhealthy fascia.

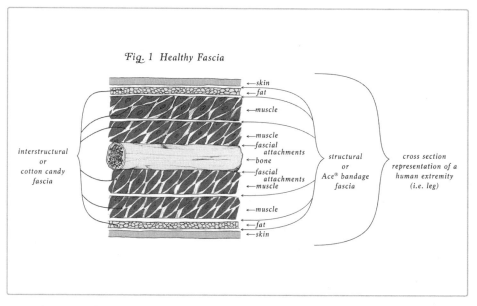

This image shows healthy fascia, which is also known as fascia that has been restored. Note smooth surface of the skin and even distribution of the fascia.

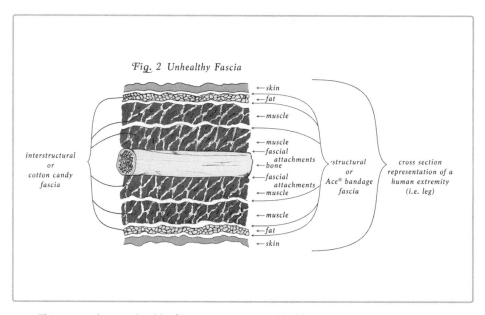

This image shows unhealthy fascia. Notice the crinkled fascia, causing discrepancies in the skin and health problems deeper in the body.

PHASES OF FASCIA FREAK-OUT

On the next few pages is a fairly comprehensive chart that describes the different stages of fascia health called the "Phases of Fascia Freak-out." There are 7 main phases with phase number 1 being very healthy and phase number 7 being the worst possible state, a full blown fascia freak-out! The chart describes what is happening in the body in each phase in the blood, nerves, muscles, spine and joints, and in your brain.

•Blood: The blood runs through the fascia. Therefore, if the fascia is unhealthy, then the blood does not flow properly.

•Nerves: Nerves also run through the fascia. Therefore, if the fascia is unhealthy the nerve signals can be impeded and the biofeedback (messages to and from the brain) can be negatively affected.

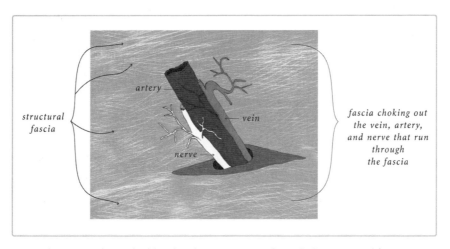

This image shows the blood and nerve running through the Structural fascia.

•Muscles: The muscles and tendons are both covered and penetrated by fascia, so unhealthy fascia can inhibit their ability to contract, relax and stretch. Muscle and tendon function is so important because their role is to provide strength and movement.

•Spine and Joints: The spine and joints have fascia running all around and through them. In fact, the spine has all 4 types of fascia affecting it. When fascia is unhealthy, the joints and spine can be compressed, twisted, or contorted.

•Brain: Last but not least, in the chart, we address the brain and overall sense of well being. Have you ever seen a human brain? It is covered in fascia! Unhealthy fascia can clamp down on our brains, causing headaches, brain fog, and a host of other symptoms.

Now, if you are only concerned about cellulite, you only need to focus on what is happening in the blood and nerves. The blood is important for the actual skin health, which will improve the overall appearance of cellulite. The nerves are important to cellulite, because the messages delivered from the brain to the body and back are a primary cause of fascia becoming unhealthy, which causes the appearance of cellulite to worsen. When properly addressed, the fascia, the blood, and nerves alone can end cellulite.

For now I want to give the "bigger picture" and I think you'll be interested to see how fascia affects everything! For this book, I could have omitted the other sections of the Phases of Fascia Freak-out Chart, but after observing the daily evolution of fascia understanding among the members of my private Facebook® group, I felt it was important to convey how fascia affects the other systems of the body as well. But because this is a book about cellulite, at the end of the chapter I will go deeper into the fascial affects on the blood and nerves and how they manifest in the skin.

With this knowledge, you can begin to interpret your own body with a higher level of understanding and connect your own dots on the picture of your health. Once you understand how you got your fascia "score" (1-7), you can understand how to move the number closer and closer to a one. For example, even with all my orthopedic and systemic issues, I can hover between a 2 and a 4 depending on how much effort I put into my fascia care.

In summary, this chart is simply about understanding "what is happening" so you can take your knowledge and the tools in the coming chapters and have control over your fascia, and hence, everything else!

Is the chart all inclusive? Absolutely not. But it is a great "quick glance" of symptoms to help you determine how healthy or unhealthy your fascia is. As you read through the stages of the chart and understand the condition of the fascia, and the symptoms, just know that you are NOT STUCK in any phase. You are now in the driver's seat to change your body! Strap in.

Phases of Fascia Freak-out

Phase 1

People in this phase are usually pro-athletes, top fitness trainers, or at the highest of the "fit" food chain. Their fascia is healthy, pliable, and functional.

CONDITION OF THE FASCIA

- Fascia and other soft tissues are healthy and hydrated.
- Body is in proper alignment.
- Proper muscles are used to perform specific movements.
- No restrictions.
- Fascia is not adhesed.
- Fascia glides easily over the muscles.
- Fascia functions properly and is able to stretch and contract.
- Fascia as a full-body entity is not sticking, distorting, or recoiling anywhere.

SIGNS/SYMPTOMS THAT PRESENT AND DIAGNOSIS

- Increased Speed
- Increased Agility
- Peak Performance
- Energy is Strong
- No Pain
- Quality Sleep
- Healthy Mental State

- Skin has Healthy Glow
- Void of Inflammation
- Feelings of Youthfulness
- Overall Structurally Sound
- Better Access to Natural Athletic Talent
- FEELS GREAT!

EXAMPLE OF PHASE 1 HEALTHY FASCIA

- Fascia is supple
- Fascia is pliable
- Fascia is hydrated
- No visible distortions
- No dents
- No stripes
- No Hail Damage

System of Body	What's Happening in the Body
Blood	• Good circulation. • Nutrients and oxygen are being properly delivered by the blood to the cells, fighting off disease and ridding the body of toxins. • During workouts, the blood flushes to muscle to achieve a training affect. • Blood is circulating back and forth through the body warding off the signs of aging.
Nerves	• Brain and nervous system have the ability to send a signal to every part of the body. • Body communicates back to the brain that everything is good. • Recoil and distortions nonexistent. • Nerves message to the fascia system is "Keep functioning properly—fascia is healthy." • Fascia is also a communicating system and is sending positive message to and from the brain.
Muscle / Tendon	• Muscles are fully accessible, from origin to insertion. • Muscles inside your joints are utilized and strengthened throughout movement, supporting their function to stabilize joints and prevent pain and injury. • Bellies of muscles are able to contract, relax, and stretch—which means the muscle is able to function and grow and be utilized as part of movement.
Spine / Joints	• Support structures are healthy—spine, discs, labrums, meniscus, ligaments, etc. • Joints are healthy and functional. • Blood is flowing through the joints, preventing inflammation. • In the spine, nerve activity to the multifidus (the muscle group that helps stabilize vertebral segments). • Blood flow around the spine and in the Spinal cord fascia. This keeps the spinal cord open and supports proper nerve activity. • Blood/fascia nerve activity nourishes discs.
Brain / Sense of Wellness	• Brain is affected by fascia because it is covered in it and penetrated by it! • Every upside that the rest of the body is experiencing in this stage, the brain is experiencing as well. Healthy nerve activity means strong signals to and from the brain. • Endorphins are more easily released. • Serotonin levels are more balanced. • Better Memory • Reduced Anxiety • Better Sleep • Better Sense of Well-being • More Energy, etc.

Phases of Fascia Freak-out

Phase 2

Something is brewing—even though people in this phase would not identify themselves as "AT RISK"—symptoms of earliest onset of fascia recoil are on the horizon. Most people in phase 2 are highly fit and only struggle with mild setbacks or 1-2 "trouble spots."

CONDITION OF THE FASCIA

- No pain in "normal circumstances."
- Mild discomforts when you "push it."
- Micro-compensations have begun, although it's unfelt.
- Structural changes start to occur.
- Micro-misalignments are present.
- Sends signal between brain and body that something is not right.

SIGNS/SYMPTOMS THAT PRESENT AND DIAGNOSIS

- Micro curve in the spine causes rare flare-ups.
- Shoulder(s) slightly up causing tension.
- Feet slightly turned out, supinated/pronated and occasional discomfort when changing shoes.
- Tiny postural problems that most have felt their whole life.

- Chiropractor/PT might tell you your pelvis is off or something with pronation/supination.
- Might have brief tendonitis spell.
- Brief plantar fasciitis episodes.
- Slight decrease in performance possible.

This is a Before and After

Fascia is primarily supple • Small dents or ripples begin • You may be cellulite free, except one spot • Cellulite may look like a small stripe(s).

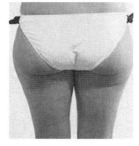

Before

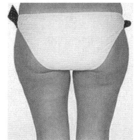

After

System of Body	What's Happening in the Body
Blood	• Minor restrictions in blood flow. • May have mild or temporary swelling after increase in activity. • May be microscopic sites where blood is restricted in micro-fascial adhesions. • Earliest onset of changing the neuro-muscular firing patterns for movement.
Nerves	• Stops firing signals where there might be a micro problem. • Starts recruiting compensatory muscles.
Muscle / Tendon	• Muscle memory starts changing. • Slight shift in center of gravity. • Some muscles are not firing fully. • Some muscles are over-firing. • May not have physical pain but starting to feel "tight." • Feels an ache and takes a painkiller. • First adhesions between muscles occur, like IT bands. • First "knots" become present.
Spine / Joints	• Joints are looking for a new function around the micro-structural deficiencies. • Not functioning optimally though still functioning. • Micro shifts and rotations within the vertebral column. • Some joints may be jammed, like hips or toes, and feel like they need to be adjusted.
Brain / Sense of Wellness	• May have occasional headaches. • Earliest stages of mental processes are starting to shift focus toward acknowledgement of discomfort. • Earliest onset of "brain fog." • Head may not ache, but neck, traps, and upper back are "knotty" and tight.

Phases of Fascia Freak-out

Phase 3

The majority of people are in this phase. People only hurt "when" or it's just a matter of "when" something is going to manifest! (e.g., I only hurt when I . . . run, lift heavy, play racquetball, etc.)

CONDITION OF THE FASCIA

- Fascia is beginning to react to small structural deficiencies.
- Pain occurs when the body is pushed.
- Fascia R.A.D. is more visible.
- Fascia is distorting the skeleton causing asymmetries of the posture to be visible.

SIGNS/SYMPTOMS THAT PRESENT AND DIAGNOSIS

May have anything from Phase 2 and:

- Probably have to ice after working out.
- Bulging disks, rib flares, shoulder off, and/or knee off.
- We can see a rotated hip, an elevated shoulder and/or curve in the spine is wrong.
- May have had plantar fasciitis.
- Early onsets of tendonitis.
- Traps are always tight.
- If the Iliotibial (IT) Bands are not rolled out—knee(s) hurt.

- If a brace is worn it doesn't hurt.
- Tension Headaches
- Mild Arthritis
- May have had a back strain.
- Joint Swelling
- Muscle Spasms and Cramps
- Orthopedics often prescribed.
- Many diagnoses like tennis elbow, runners knee, plantar fasciitis, hip pointer, or degenerative disc are being handed out.

This is a Before and After

Fascia is re-coiling • Cellulite is present • Body feels tight • Skin can be pinched, but some places are chunks.

Before   After

System of Body	What's Happening in the Body
Blood	• Fascia restrictions prevent blood flow to the areas that need it most. • Blood is pooling in the compensatory muscles causing swelling in the joints (wherever the body goes to cheat is where it swells). • Small spider and varicose veins.
Nerves	• Nerve signal cannot get through the fascia to the proper muscle. • Brain says via the nerves, "That's going to hurt — don't use it!" • Nervous system changes the muscle memory for entire body. • Dramatic shifts in center of gravity. • Muscle memory is no longer efficient and workouts hurt vs. feel better at least some of the time.
Muscle / Tendon	• When training, severe lactic acid builds up resulting muscle soreness. • Muscles are over-developing and under-developing. • Increase in muscle imbalance. • Muscles spasms begin. (When the muscle is used too much, it clamps down and shuts off. Eventually, you create muscles that don't ever let go—always tight.) • You may begin to wear a brace/Ace® Bandage.
Spine / Joints	• Spine is starting to change. • Sections of the spine doing too much; others not doing enough causes curves and scoliosis. • Fascia crossing the joints tighten—closing the joints a little bit all over. • Range of motion is restricted. • Swelling or inflammation may be present.
Brain / Sense of Wellness	• Start storing stress in places of imbalance in the physical body. • When feeling stress, compensatory muscles are over-contracted—worsening the condition. • Mental capacity is diverted to pain. • Body is sending a message to the brain, "I'm not doing well all over—help me out." • Brain has to process "I'm not doing well," which further deteriorates mental faculties.

Phases of Fascia Freak-out

Phase 4

Fascia system in SERIOUS TROUBLE! Most people begin to get hurt every time they exercise.

CONDITION OF THE FASCIA

- Fascia is adhering in the joints.
- Fascia is puckering and pinching all over.
- Regular discomfort.
- Spurts of pain.

- Fascia is pulling like a "tug-of-war" between structures causing major dysfunction.
- Surface adhesions.
- Muscle-to-muscle adhesions.
- Interstructural adhesions.

SIGNS/SYMPTOMS THAT PRESENT AND DIAGNOSIS

May have anything from Phases 2-3 and:

- Many are labeled with fibromyalgia.
- At this point, most have consulted professional help.
- Wrist has hurt for 3 weeks straight.
- Have back pain, blow it out, it gets better, blow it out again.
- Mystery pains that come and go.
- In your back—herniations, bulges, joint impingement.
- In your neck—headaches and TMJ.

- Plantar fasciitis—achy, throbbing feeling.
- Radiating pain from original site of injury/dysfunction.
- A "little" arthritis.
- Holding weight (fat) in a specific area.
- Random cellulite appears.
- Muscle strains or injury are more prevalent.

This is a Before and After

This is usually the ugliest cellulite stage • Hail Damage dents & Gummy Bear can be present (see AshleyBlackGuru.com/TheCelluliteMyth) • Fascia is grabbing muscles and fat, making you look "chunky"

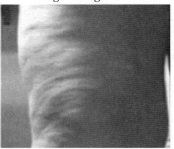

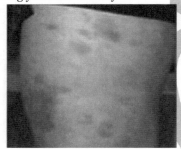

Before   After

System of Body	What's Happening in the Body
Blood	• Serious restriction of blood flow due to fascia adhesions. • Blood will either completely bypass the area, making it feel numb, or it will pool around the area making it swell (e.g., pooling in the knee equals numb feet). • Common to get diagnosed with peripheral vascular disease. • Varicose veins popping up.
Nerves	• Nerve pathways go COMPLETELY around the affected area and shut off the muscle activity. • Body learns to function dysfunctionally. • More limited to specific site areas versus entire body. • Pinched nerves common. • Bouts of sciatica.
Muscle / Tendon	• You are significantly over-training the compensatory muscles. • Muscles needed for proper form or activity are virtually off. • Brain senses issues and begins to reroute around dysfunction. • Muscles can't condition effectively. • Muscles overfire. • Tendons twist/jam joints. • Chronic IT bands.
Spine / Joints	• Further joint compression occurs. • Neck is becoming flat. • Ribs are starting to get encased in fascia and lose mobility. • Discs are dehydrating, bulging and more disc damage is on the horizon. • Spine is taking the brunt of imbalances existing in other joints. • Scoliosis is visible. • Lordosis is a problem. • Kyphosis is a problem. • Jammed hips. • Joint swelling is prevalent.
Brain / Sense of Wellness	• Beginning to feel tired on regular basis. • Disruption in the thought process. • More negative energy devoted to pain. • Message from the body to the brain is, "We are messed up, so we need to change things up a bit." • By this phase, you are changing up your routine; it's affecting your workout. • May not be in pain, but probably have a disc issue or a "nagging thing" in your body. • After activity, takes a muscle relaxer. • May be on a regular dose of an anti-inflammatory.

Phases of Fascia Freak-out

Phase 5

Fascia disrupts quality of life. Exercise is a real struggle. These people have given up workouts, or they are trying other rehabilitation programs.

CONDITION OF THE FASCIA

- The fascia is now strangling the tendons.
- Palpation of tendons feel like rock hard cords.
- Fascia is clamping like a boa constrictor.
- Severely limited range of motion.
- Probably 5 or more places on the body where there is serious risk for major injury.
- Over-training.

SIGNS/SYMPTOMS THAT PRESENT AND DIAGNOSIS

May have anything from Phases 2-4 and:

- People are MISDIAGNOSED at stage 5 because symptoms seem unrelated.
- People begin to get diagnosed with mixed tissue disease, fibromyalgia, lupus, chronic fatigue.
- Crohn's disease can develop.
- Chronic Migraines
- Chronic Tendonitis
- Bursitis
- Arthritis
- May have chronic swelling visible.

- A joint that doesn't "look right."
- Restless Leg Syndrome starts here.
- Any type of chronic pain diagnosis may be getting chronic.
- Notable rib shifts.
- Scoliosis
- Hump at the base of the neck.
- Something significantly visible on an MRI, but not always.
- Bone spurs, micro tears, bursitis, and/or chronic inflammation.

This is a Before and After

All 4 types present less severe • few cases.

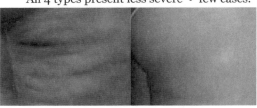

Before   After

System of Body	What's Happening in the Body
Blood	• Flow is severely derailed in specific spots of the body. • Limbs have a feeling of falling asleep. • People say, "This feels heavy" when they train (which is lack of blood flow). • Skin begins to look "dead" or older than the individual.
Nerves	• There is now a total re-routing of signals to the muscle and MAJOR shifts in gravity. • This re-routing is causing major exhaustion. • The nerves are working extreme overtime trying to fire though a clamping web of fascia.
Muscle / Tendon	• More extreme compensations to the point that there is SEVERE muscle imbalance in the joint. • Unused muscles are now completely atrophied. • Compensatory muscles are now in a constant state of spasm. • Places in the body that feel like tight cords. • Around the muscle, the fascia is trying to protect and stop the deterioration. • Inside the muscle, the fascia continues to tighten.
Spine / Joints	• Fascia is changing joints & spine causing more severe compression and dysfunction in specific areas. • Multiple bulges. • Herniation, rupture or bulging of the discs is possible. • May be developing spondylitis. • Once in later phases you start to enter the point of no return and it becomes more difficult to correct the fascia system.
Brain / Sense of Wellness	• Continuation of level 4 but more severe. • Having thoughts of pain as much as every hour. • Both subconscious and conscious awareness. • Not just altering activities but now needing to discontinue activities. • Instinctively addressing the pain. • Acknowledging that something is wrong. • Considering a surgery. • Occasional ringing ears. • Blurred vision.

Phases of Fascia Freak-out

Phase 6

Fascia is winning the battle.

They feel like they've "tried everything." They become hopeless.
Other treatments become less and less effective.

CONDITION OF THE FASCIA

- Fascia is in FULL BLOWN recoil.
- If you reach down to pinch the skin away from the muscle, you can't pull it up.
- The fascia is like a spacebag—clamping from the outside-in.
- Inflammation is now trapped in tight fascia all over the body.
- Multiple sites of adhesions—severe disruption in the overall fascia web.
- Entire fascia system is saying, "We need to help!"

SIGNS/SYMPTOMS THAT PRESENT AND DIAGNOSIS

May have anything from Phases 2-5 and:

- Fibromyalgia is "go-to" diagnosis
- Chronic Fatigue Syndrome
- Migraines
- Shin Splints
- Spondylitis, spondylosis, all the "spondys"
- Numbness and shooting pain in multiple places

- Neuromas in the feet and hands
- Chronic Inflammatory Disease
- Lipedma
- Lyme Disease
- Hypothyroid
- Dystonia
- Scleroderma

This is a Before and After

All 4 types present severely • Early stages of Beyond Bound™

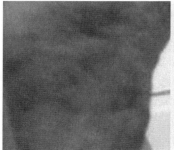 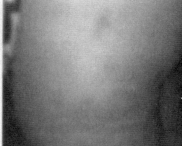

Before   After

System of Body	What's Happening in the Body
Blood	• Severely restricted, pooling in different parts of the joints. • Full-body circulation is a challenge. May be diagnosed with high blood pressure as a result. • May have chronic swelling or numbness in an extremity.
Nerves	• Nerve struggles to signal for basic movement—body is just trying to hang on. • Probably have some sort of altered gait, a limp, a drop foot, or an arm that doesn't swing. • Can't move and exercise. • Nerve pattern is so dysfunctional that muscles are just wasting away.
Muscle / Tendon	• Body can no longer stabilize. • Not getting adequate blood or nerve supply, atrophying and wasting away. • Tendons are non-functioning and recruiting bellies of muscles to perform joint actions.
Spine / Joints	• Most likely there is constant pain in the upper, mid, or lower, and likely all three. • The fascia is strangling the body of the inner structures. • Spondylitis is usually present. • Severe SI dysfunction.
Brain / Sense of Wellness	• Discontinuing physical activity—completely changing your life around the pain. • Body is freaking out all over. • Sleep is affected. • Skin becomes sensitive to the touch. • Any strenuous activity can make one bedridden for a day or two. • Even simple tasks, like sustained walking, can cause a fascia freak-out. • May be on pain killers. • May be medicating for migraines. • Significant problems with mental focus. • Pain wears on the face. • Ringing in ears. • Severe problems with eye opening. • Can't get the energy to start. • Center of gravity is severely altered and visible. • Balance is challenged.

Phases of Fascia Freak-out

Phase 7

Fascia intervention is of utmost importance.
These people are basically revolving their lives around their pain.

CONDITION OF THE FASCIA

- Full Blown Fascia Freak-out!
- Fascia is so locked down—chronically in pain all over.
- Everything feels hot.
- Can't be touched.

SIGNS/SYMPTOMS THAT PRESENT AND DIAGNOSIS

May have anything from Phases 2-6 and:

- Painful full-body sensitivity.
- Back pain raging.
- Nothing is comfortable.
- Hot to the touch.
- Can't sleep.

- Most considered as "mystery" cases.
- Anything chronic .
- Every diagnosis in the book related to "symptoms" is given.
- Usually referred to pain management specialist.

This is a Before and After
All 4 types can be present. • Most likely Beyond Bound™.
• Can't pinch skin. • Body is one solid chunk.
• Skin is painful.
• Fascial total disaster IMMEDIATE intervention needed

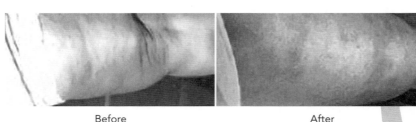

Before   After

System of Body	What's Happening in the Body
Blood	• Extremely difficult to pump blood through the body and exhausting. • Probably have a heart condition.
Nerves	• Systems of the body are shutting down. • Fascia system has had enough and is not communicating.
Muscle / Tendon	• Minimal functioning. • Entire body muscular and soft tissue tension. • Muscle movement impaired and painful. • Significant strength and flexibility deficits.
Spine / Joints	• Fascia is so tight that joints are completely jammed. • Bones are starting to fuse. • These people have sought surgery or already had multiple.
Brain / Sense of Wellness	• Confined to a bed and taking pain pills. • Entire life revolves around pain. • Popping 5-6-7 painkillers a day and can't get out of pain. • Migraines out of control/vomiting. • Hopelessness. • ZERO energy.

Now that you can see how the condition of the fascia affects the condition of everything in the body, let's go back and explore what is happening in the blood and nerves a little more, since they are the primary influences on the appearance of your skin and cellulite.

Remember our duct tape visual? Imagine a long piece of duct tape laid out smoothly with a drinking straw attached as it lies flat. (Any plastic drinking straw will do. Don't overthink it.) Now, imagine bunching up the duct tape into a sticky, wadded, balled up mess. How efficiently is ANYTHING going to flow through that straw with it all bent, mangled, contorted and compressed by the duct tape? Similarly, your nerves and arteries become compressed and restricted (or partially restricted) within the sticky, "balled up" unhealthy fascia, which naturally affects the condition and appearance of your skin. Here's how . . .

THE SKINNY ON YOUR SKIN

Keeping the skin healthy and happy requires nutrients that are supplied by the blood via a sophisticated internal delivery system, sort of like the root system of a tree but in reverse. The "roots" which are part of the circulatory system, come up through the fascia in order to reach the skin. The larger "roots" are deeper down and the smaller off-shoots, about the thickness of a strand of hair, come closer to the surface and push through the layers of fascia to deliver oxygen and nutrients to the skin cells. This delivery system also delivers the building blocks of collagen, which is the "holy grail" of beautiful skin! Collagen makes the skin healthy, tight, smooth, and youthful; and yes, it is produced internally, not externally or topically. Also, elastin, which is part of our DNA at birth, needs oxygen delivered via the blood in order to thrive at the skin level. Elastin helps the skin keep its original shape and without proper blood flow, the elastin "starves," so to speak, and can't do its job properly.

Another interesting thing about the skin is that the tiny tubes of the blood vessels also contribute to the regulation of body heat by controlling the amount of blood flow to the surface of the skin. Notice how a person with poor circulation is often cold? This also indicates that the fascia is unhealthy and the skin isn't getting what it needs because blood flow is restricted. You've probably never heard someone say, "I'm freezing! I really need to address my tight fascia that's restricting blood flow!" However, tight fascia can absolutely affect circulation, which affects the health of the skin and as I've said before . . . everything else in the body. When fascia is unhealthy due to R.A.D., it clamps down on those tiny blood vessels and restricts blood flow like a kink in a garden hose. Ever wonder why some people's limbs go numb or feel tingly? Yep. Restricted blood flow could be the culprit. How is this going to affect your skin? When the skin doesn't get the blood, oxygen, and nutrients that it needs, it's going to look droopy, saggy, wrinkly, dimply, and of course, it will be cold. If you are trying to affect the condition of the skin from the outside with creams and lotions, it's important to realize that there's very little that you can do without proper blood flow on the inside. Fascia isn't the only reason that blood flow can be restricted, but it's certainly a key player, and most people aren't even considering it! By and large, our society keeps attacking the symptoms of problems instead of getting to the root of the problems. That's why fascial restoration has such dramatic results on things like crepe skin, coloration, droopy skin, aging skin, and overall skin health—because fascia is at the root of the problem! (Literally.) Check out my website and FasciaBlasters™ official Facebook® page to see thousands of pictures that show how restoring the fascia changes appearance!

Now that you understand how the condition of the fascia affects blood flow to the skin, imagine how it affects blood flow to every other part of your body? (**#LightBulbMoment**) If blood flow and nerve activity can be disrupted by unhealthy fascia, then the functionality of any system in the body can be disrupted by unhealthy fascia.

BRAIN TO BODY CONNECTION: THE NERVOUS SYSTEM

Your fascia and your brain are in constant communication via the nervous system housed in your spine. Think of the brain as command central and the fascia is the "man on the streets." The fascia is reporting to the brain about what's going on in the body: "The hip is unstable." "The shoulder is turned forward." "Something is not right out here!" The brain then assimilates the information and sends instructions back to the fascia: "Tighten down around the hip to stabilize." "Hold the shoulder in place so it doesn't move further forward and cause injury." "Brace the back for impact from that improper foot-strike!"

This is a very simplified presentation of the proprioception of the fascia, but just know there is constant communication about the way you stand, sit, bend, move, strike your foot on the ground, everything.

When there are restrictions, adhesions, and distortions (R.A.D.) of any kind present in the fascia, the brain sees these as injuries and reacts by using the entire fascia system as a protective mechanism. Protection comes in the form of a lockdown and when the fascia locks down, as you saw in the "Phases of Fascia Freak-out Chart," the effect on the body is dramatic! It's your body's way of screaming at you, "Stop what you are doing and pay attention to me!" The body screams at the brain, then the brain screams back at the body. (And you've probably felt your body screaming at some point.) This ill mind/body connection caused by structural deficiencies and unhealthy fascia can create a vicious, self-perpetuating cycle that can keep you in a constant state of lockdown, pain, tightness, and cellulite. More on this when we talk about structural deficiencies coming up and also what you can do to fix all of it. Take heart, there is a real solution!

Now, let's get back to blood flow and nerve activity. As I've explained, the blood and nerves literally run side by side within the fascia. Nerves are responsible for telling muscles and other systems how to work. Imagine the implications if nerve activity and blood flow are impeded because of unhealthy fascia or R.A.D. The effect is dramatic! This is where I might need to get out my soap box. Here it comes, ladies! *Remember, fascia surrounds, penetrates, and interplays with everything in the body, meaning that it affects and is affected by whatever it touches.* Because of this, the condition of the fascia directly affects the condition of everything else in the body, not just your cellulite. So, the question is, do you really have cellulite or unhealthy fascia? Do you have a back problem or unhealthy fascia? Do you have premature aging or unhealthy fascia? Do you have migraines or unhealthy fascia? Do you have a liver problem or unhealthy fascia? Do you have TMJ or unhealthy fascia? Do you have scar tissue or unhealthy fascia?

I will let you play the "Wow, what if restoring the fascia could . . . game," just as I do every night when my head hits the pillow. The implications are almost unfathomable. The results our users are reporting are expanding to conditions that even I have never thought of!

I'm not saying that restoring the fascia will cure every disease and give you a perfect, lean body. I'm not even saying that you won't ever have cellulite again because you can trigger unhealthy fascia, but you'll know what to do about it. All I am saying right now is that the condition of the fascia affects everything in the body and as a society we need to put fascia on the table as one of the primary focuses and start exploring it more. The conversation needs to be started, the research needs to be conducted, and it's time that Fasciology takes center stage . . . especially for cellulite, because it is undeniably the cause! (Yes, call the Nobel Prize® committee . . . your search—*the* search—is over.)

Jennifer Scott
June 19

My 8 year old daughter just thanked God for Ashley black while praying over our dinner. She said, "Thank you for the people who gave my mom the tools that are helping her get better." I have had daily chronic headaches, TMJ pain, neck pain, and back pain for years. Ashley's protocols are really helping me feel looser and better. Our family has prayed daily that God would heal me from my debilitating pain and He has answered my prayer with your motivation to help others, Ashley! My daughter's prayer brought tears to my eyes. I am not 100 percent but everyday I feel better and better! I have only been applying the protocols for 2 weeks but I work on some areas every day. Thank you for giving my family their mommy and wife back! I was so desperate for pain relief!

4

Why Fascia Becomes Unhealthy

"The scientist is not the person who gives the right answers, but the one who asks the right questions."

—Claude Lévi-Strauss

Everyone wants to know why fascia gets "angry" or becomes unhealthy, and that's a good question to ask. Truthfully, there are lots of reasons. Not only that, the reasons are different for every person. It's sort of like asking why a person has gained 10 pounds, or why their hand hurts, or why their posture has changed. It could be lots of things. Food, stress, hormones, allergies, lifestyle choices, schedule inconsistencies, and so on. Regarding cellulite, some schools of thought say it's due to hormones, smoking, diet and exercise, or biomechanics; and we know that genetics and hormones do play a role, but there's so much more—toxins trapped in the tissue, meds, hydration, your micro biome, and more. Remember, fascia is a system made of up of cells so *anything* that affects cellular health affects your fascia!

The exact relationship of all these factors is part of an evolving science, but we do know there is a lot that impacts the fascia. This is why there are many cellulite "treatments" that work in varying degrees, which we will talk about shortly, because they address different stressors to the fascia system. Some treatments might have an impact for some people but not for others depending on each person's root cause for fascial restriction, adhesion, and distortion (R.A.D.). At the end of the day, because cellulite is ultimately a fascia problem, everyone who restores their fascia will see results regardless of their root causes!

There are, however, three top factors that I consistently see that negatively impact the fascia the most, they are: crappy structure and biomechanics, dehydration, and blood flow. This top three list comes from personal experience and is 100% my opinion, but I think it's as solid as any other theory out there, and it's my book. (Wink.)

1. CRAPPY STRUCTURE & CRAPPY BIOMECHANICS

Two things that will jack up fascia in a jiffy are when you have a crappy structure, and then when you move that crappy structure, well . . . crappily. Let me break it down for you. Your "structure" is simply the condition of your physical human body. If you have a problem in your hip, you have what is called a structural deficiency in your hip. If you have carpel tunnel, you have a structural deficiency in your forearms. If you have poor posture or discs out of place, you have structural malalignments. If your body is riddled with R.A.D., then your structure is fascially crappy and crappy fascia will affect your biomechanics, which is a fancy word for "the way you move your structure." Just to recap, structure is your physical body; biomechanics is the way you move your structure.

In a perfect human structure: every joint would be fully open, every inch of fascia would be supple, every ligament intact, every bone healthy, scar tissue broken down, every nerve free to fire, every muscle fiber accessible, every blood vessel open, and every bony landmark in alignment. We have many pictures of changed structures on our

Facebook® page and at AshleyBlackGuru.com/TheCelluliteMyth. Have you ever thought about how symmetry and beauty go hand in hand? Research shows that babies tend to stare more at what we would call beautiful faces, but what they see is symmetry, and symmetry is easier for the eyes and brain to process than when things are not symmetrical. Remember our brain to body connection? A symmetrical structure is also easier for the brain to communicate with. So when you hear words like "alignment," "symmetry," "posture," and so forth, these are all related to what I call structural integrity.

In a perfect biomechanically functioning human, the nerves would fire each movement through the most efficient muscle movement patterns, no compensation would occur, every single bone would remain in alignment throughout all sports and daily activities, and every movement would be performed with a perfect, balanced center of gravity. *If* perfect structural and mechanical integrity could occur, there would be NO repetitive movement orthopedic injuries or pain, and many neurological, blood, and toxicity ailments would be minimized. And of course, there would be no appearance of cellulite. That's a bold statement, but it is absolutely TRUE. So the goal of body care should always be to get as close to structurally and biomechanically perfect as possible. **#Preach #AshleyKnowsBest**. Even if you shoot for perfect and miss, you will be far better off than not addressing these at all.

Having said that, I'm going to guess that no one has ever told you this but there is one way, and one way only, to stand properly. There is one way, and one way only, to sit properly. There is one way, and one way only, to walk properly. There is one way, and one way only, to run properly. There is a proper way to perform any movement and surprisingly, the knowledge is lost among most people today. Even at the highest level of pro sports; I know because I worked in this arena for 15 plus years, and in gyms and doctors' offices, they don't teach you how to stand properly, or sit properly.

Crappy biomechanics, or not moving the right way, can be the byproduct of many different causes such as your medical/injury history, environment, or genetics, but it is most often caused by small malalignments in the body exasperated over time. For example, do you lean slightly to the left when you walk or run? Do you have a foot turned out when you squat? Do you wear shoes with heels that are too high? Do you sit at your desk with a hump in your back? Is your neck bent and compressed when you play with your kids? There is a proper way to do all of those movements. Amazingly, this is rarely addressed or, at the very least, improperly addressed. Moving the proper way, or with the proper biomechanics, minimizes stress on the body and ultimately the fascia system. If you are out of proper alignment every time you take a step, and you do that 50,000 times a week, there is going to be a dramatic impact on your brain's message to your body. Every single thing about the way you hold your body affects the fascia. Is a hip rotated? Shoulder raised? Neck forward? Yep! Tight fascia is right on the scene with it.

There's always more research needed, but in my experience, poor posture and biomechanics are the number one cause of unhealthy fascia or R.A.D. Sometimes the structure is stuck by R.A.D., which causes poor biomechanics and sometimes the biomechanics consist of poor movement patterns, which cause the structure to change. Like a bad two-way street, they affect one another. Crappy biomechanics perpetuates crappy structure, and sometimes crappy structure perpetuates crappy biomechanics. When this happens over and over and over again, several things happen: we get aches and pains, our workouts are less effective, other systems are affected, our overall appearance suffers and in time, this is how we age improperly. That's why it's so important to change your structural integrity AND your biomechanical integrity. When you do, magic will happen!

Now, in an extreme enough situation, the body can go into total body lockdown, like Phase 7 on the chart, which always goes back to alignment and symmetry. A lockdown can cause a total body malalignment. When the body is out of alignment, the brain starts making compensations again and adjusts itself slightly and continuously. Think of a game of Jenga® where you have a tower of small wooden pegs. One person pulls a peg from the bottom and has to balance the tower at the top. One wrong move and the entire tower will fall over. This is sort of what the body is doing. You have to constantly redistribute the weight or you are going to have a problem. Now, if you pull out a roll of duct tape and wrap it around that Jenga® tower, it would stabilize the structure, which is what the brain tells the body to do with the fascia. The fascia clamps down around the joints and weak muscles like duct tape to offer some extra support and protect the body from injury as much as possible. While this sounds like a clever stopgap, the tight fascia restricts movement further and a giant snowball of problems ensues. More compensation. More clamping down. More tightness. More pain. More restriction, and so on. Joints are stressed, ligaments are stressed, muscles are doing things they weren't designed to do, blood flow and nerve activity are restricted, and well, you just keep feeling worse and worse. And don't forget the whole reason you bought this book—yep! You guessed it, fascia restrictions, adhesions, and distortions (R.A.D.) that look just like what we call cellulite, which is exactly why I call it a myth! It's not something in and of itself; it's the appearance of fat popping through the unhealthy fascia.

#CanIGetanAmen

In the most simplistic approach, you can improve your structural integrity with manual fascia treatment (as outlined in chapter 8), stretching, and remodeling of the fascia. Mechanical integrity can be improved with proper exercise mechanics and movement pattern awareness. There's a whole book waiting to be written just on this topic alone, but check out my website for resources. I also have an entire playlist on YouTube® where we analyze people's structure. What's interesting is that every single person featured on the playlist, no matter age or level of health or fitness, has deficiencies in the way that they stand. And if you can't stand properly, how in the world are you going to lift weights properly, or do a boot camp class, or pilates, or lift kettle bells, or yoga, or run, or anything else? You can't. Every single movement will cause varying degrees of damage if your stationary structure is out of alignment.

In the real world, having structural perfection and mechanical perfection would indeed be rare and most likely impossible. However, Fasciology is a portal to transform your current beliefs about your physical well-being into a new paradigm where the "unattainable" becomes ATTAINABLE! This is how I have become the physical manifestation of how powerful this is. I should be in a wheelchair—and instead I'm fit and active, I heli-ski, do hot yoga, dance, and I'm a hardcore surfer. (My story is in the last chapter.) I am extreme, and maybe you are too . . . but my fascial health, which determines my structural and mechanical integrity, is the difference from being in crippling and blinding pain and being an athletic, 44-year-old, world traveler!

2. DEHYDRATION

The second greatest impact on the health of the fascia tissue is dehydration. When most people think of being dehydrated, they think, "Oh, it's hot outside. Better drink some water so I don't pass out." But hydration is so much more than just preventing overheating. Hydration keeps the fascia functioning properly and is crucial to help you move without pain.

The earliest phases of dehydration are sometimes undetectable, or manifest as aches and pains, particularly a headache. The person experiencing the random aches and pains will often take a painkiller instead of a drink of water, and the problem escalates. You may not even be thirsty and be dehydrated because thirst is actually a more advanced sign of dehydration. Here's how the slightest bit of dehydration affects your fascia, your mobility, metabolism, and the appearance of your cellulite!

The fascia is connected to whatever it surrounds and penetrates in the body all the way down to the cellular level. It's not like bread touching meat in a sandwich; it's more complex than that. It's more like the mayonnaise in the chicken salad. The teeny tiny cells are commingled. So fascia isn't lying on top of muscle fiber, fascia cells are literally commingled with the muscle fiber cells. I really want you to get a picture of how this works in your body, because you may be dehydrated right now and not even realize it. Staying well hydrated is a piece of the cellulite puzzle!

Cells are round and when they touch there is still space above and below the area that is touching. Just imagine four tennis balls crammed into a small box. They don't fit exactly; there is still some space in the box. This space surrounding the spots where the cells touch is called the extracellular matrix. (Just like the movie!) This extracellular matrix isn't empty space; it is filled with fluid that flows around the cells, delivering nutrients and washing over the cells, removing waste and toxins. And of course, this fluid has a very fancy name that you will never have to remember after reading this book but I'm going to give it to you anyway so you can impress your friends and family—this very important fluid is called interstitial fluid. So the interstitial fluid is flowing through your body, sprucing things up, taking out the trash, and making sure those little cells have what they need to be happy and function properly; sort of like the nurses in the maternity ward. The nurses go from baby to baby caring individually for each one, feeding them and changing diapers so the babies are happy, healthy, and growing. Now, what would happen in the maternity ward if there were a shift change and the current nurses leave and not enough new nurses show up? The babies cry, the diapers stink, and it very quickly becomes a stressful place instead of a happy, healthy place. This is what happens to your cells when there isn't enough interstitial fluid. Your cells "cry" and the fascia webbing shrivels up and sticks together like plastic wrap instead of being plump and healthy. When the fascia clamps down, you know what happens next—pain, restriction, cellulite and dents.

Keep in mind that interstitial fluid is constantly leaving your body so you need to replenish it daily by drinking ample water. The interstitial fluid flow has an effect on tissue function, cell migration, tissue differentiation, tissue remodeling and tissue morphogenesis. Again, these are more terms you don't really need to worry about unless you're just curious. The takeaway here is that the slightest variation in the amount of interstitial fluid alters the properties of the fascia tissue, which impacts the way you move. If you are properly hydrated, the cells can adapt to the stresses on the body. This basically means, if you are dehydrated and moving around, you're more likely to jack yourself up than when you are fully hydrated.

Electrolytes

Not only is the interstitial fluid crucial to hydration, but the cells have to open to receive the nutrients that are in the interstitial fluid. In other words, the baby has to open her mouth to receive the bottle and the bottle needs to be filled with the right stuff. The right stuff is electrolytes! An electrolyte is a substance that carries an electrical charge, which is essential for life. (Hence the "electro" part.) We need electrolytes to survive. Electrolytes are minerals, the 7 major electrolytes are: sodium, potassium, calcium, bicarbonate, magnesium, chloride and hydrogen phosphate. Electrolytes are responsible for balancing blood pH, hormone synthesis, muscle contraction, nerve conduction, digestion, detoxification, and of course, hydration. You can get them from fruits and vegetables or through an electrolyte supplement. Electrolytes are so important to the fascia system that as I have researched and evaluated brands, I just wasn't able to find one that met the "Ashley Black standards," so I formulated my own called, Cellectrolytes™, which you can find on my website. It's the perfect balance of electrolytes to nourish your cells and support your fascia! Of course, I always recommend my own products! If you don't use mine, make sure you supplement or get them from food, because they are a huge part of hydration and cellulite!

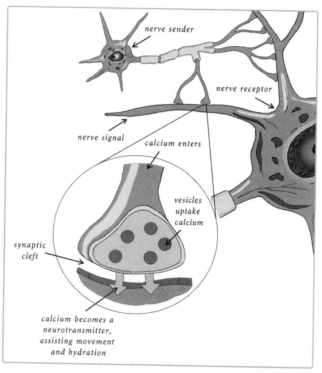

nerve sender

nerve receptor

nerve signal

calcium enters

vesicles uptake calcium

synaptic cleft

calcium becomes a neurotransmitter, assisting movement and hydration

This image shows electrolytes' contribution to nerve transmission, i.e., how movement happens.

Finally, how do you know if you are fully hydrated? Don't wait until you're thirsty to drink up. The best way to determine if you're drinking enough is by taking a look at the color of the fluid leaving your body, which should be pretty close to the color of the water entering your body. There's a lot more to be said about hydration and the body, but there are plenty of resources and books to help you learn more. The important takeaway for this book is to support your fascia by drinking plenty of water and supplementing your cells with electrolytes.

And, the next time you go to the bathroom, think, "Shift change!" And keep those little cells clean, fed, and happy! When your cells are happy, your fascia is happy, and when your fascia is happy, your skin is happy and free of cellulite! And that will make *you* happy!

3. BLOOD FLOW

The third greatest impact on the health of the fascia tissue is blood flow. There is more than one way to stimulate blood flow but cardiovascular physical activity is at the top of the list. That means elevating the heart rate to 135 to 170 beats per minute. The Centers for Disease Control and Prevention (CDC) says the fastest-rising killer in America is inactivity. Inactivity is worse than any disease. Why? Because if a part of the body isn't being stimulated, the blood won't travel to it, which means no oxygen travels to the cells and the cells begin to slowly die off. This is called ischemic death and it happens in your fascia as much as anywhere else. Blood is the life source of the body. Have you ever cared for someone in a hospital bed? If a person is immobile, the staff will put compression boots on them to stimulate blood flow throughout the body. The autonomic nervous system, which has fascial connections, (illustration on page 54) responds to the stimulation and triggers the blood flow.

When we perform exercises, the activity causes fascia angiogenesis, which is a fancy way of saying that the body senses that it needs more blood and will create new blood vessels and capillaries to adapt to the activity. When we aren't performing cardiovascular exercises, the opposite happens and the body restricts blood flow to adapt to inactivity and cells begin to die off.

Blood flow is also key to building muscle—we place the demand for blood to rush to the area when we use a muscle. If a muscle doesn't get a stimulus, it gets smaller or atrophies, which means it becomes mushy and doesn't function, or partially doesn't function. The body must be stimulated to keep the blood flowing, and when we get to the protocols, you will see my favorite ways to do this. Just make a mental note that healthy blood flow = healthy fascia = reduced appearance of cellulite!

Hopefully by now it is abundantly clear how cellulite is merely a side effect or symptom, if you will, of unhealthy, distorted fascia. Everyone has fascial distortions; they just present differently in appearance and in how they impact people. Yes, distortions show up as cellulite; however, really bad, and I mean REALLY bad fascial recoil can also show up as the illusion of smooth skin, which I will explain in chapter 6. Sometimes, because of those tiny biomechanical imbalances, a person may appear to be in the best physical condition when in reality, their fascia is in the worst kind of condition. The truth hurts sometimes but it's also the first step to healing. You may be distorted today but things can be different tomorrow! (Or in about 60 to 90 tomorrows.)

HOW DID WE "MYTH" THIS?

The information is right here in black and white but the big question is, how in the world did the rest of the planet myth this!? How are we still calling hernias "hernias" when a hernia is clearly torn fascia? How is it that WebMD® says the only treatment for plantar fasciitis is to rest the foot, when you can get out of pain almost instantly by addressing the fascial distortion instead? Why are we cutting on women who have "turkey neck" when it's just a structurally compressed and shortened neck with tight fascia causing skin gaps? Why aren't we connecting the dots about the structure of the body and the appearance and function of the body? For example, did you know that fat rolls in your back can be the result of tight fascia causing the pelvis to tilt forward, like a bucket of water spilling out the front creating fat bunches above it? **#GetItTogether**

Okay, so I get a little passionate about this. Correction, I get a *lot* passionate about this because when I owned Fasciology centers, people came to me all the time after they had no hope and in as little as one to five visits, their lives were radically changed. By now you may also be wondering why people are coming to me in droves, how in the world did I discover the cure for cellulite? Well, I'm going to tell you my death-defying story at the end of this book. (And yes, it truly is a death-defying story.)

But one of the reasons this mind-blowing, obvious fact has been missed is that many experts in body care perceive fascia as packing material, like bubble wrap, filling up space in the body. To some degree it is; however, it's so much more. Additionally, in dissection courses it is common practice to teach students to scoop the fascia out of the way. Yes, much like hand washing was once disregarded, fascia has been largely disregarded too. It's mind-boggling to me because if you stop and look at the fascia, if you hold a piece of fascia up to the light, you will see that it has blood and nerves in it, which means it is a living tissue inside the body. But it's not talked about in many of the medical textbooks, or textbooks in general. In fact, at my home I have *Body: The Complete Human* by National Geographic® and it doesn't mention anything about fascia at all. It's not even in the glossary. Even the more cutting edge health crusaders like Dr. Oz® have barely covered the subject. I nearly fell out of my chair in disbelief when I was watching his show on the ABCs of health, and fascia was NOT the "f" word. Our culture is so unbelievably underedu-cated! I would like to give honorable mention to the medical textbook *Gray's Anatomy* because it does actually talk about fascia. All of this is NOT to "call" anyone out, but just to say that "the institute" and "pop culture" haven't yet gotten on board at a mass level.

We also know that Leonardo da Vinci, one of the greatest minds of all time, studied fascia. As early as the late 1400s, he dissected and drew anatomical structures of the human body which included the fascia. He studied the mechanical function of the human skeleton and muscular system predating modern biomechanics. And if you want to know where he got these cadavers to dissect, well, there's a reason it's against the law to dig up graves. **#ThanksLeo**

Remember what I said earlier about the notion that in order to be perceived as an expert by your peers, you have to follow what everyone else before you did? There's a saying in the South, "If you always do what you always did, you'll always get what you always got." What our society has now is a completely segregated approach to every aspect of the human body—we have skin specialists and eye specialists, fitness specialists, dietitians, heart specialists, plastic surgeons, and the list goes on and on. This is excellent and precise on the one hand, but on the other hand, it's a problem because it's rare that the specialists work together to form the bigger picture. It's sort of like watching an IMAX® movie from the front row. Being close isn't always best because you only really see what's directly in front of you. Sitting farther back allows you to take everything in so you can make better sense of it all. (Sometimes I feel like I am watching several IMAX® screens from across the street when it comes to fascia!)

Similarly, cellulite people don't study fascia and fascia people don't study cellulite.

As it is, fascia experts are few and far between. I don't know of a single fascia expert who has connected the dots between fascia and cellulite. I say this because I searched for the words "fascia" and "cellulite" online, and I was the only thing that came up! By the release of this book, that may not be the case, as I'm shouting from every rooftop between now and then.

Of all the research coming out of the Fascia Research Congress™, there's not one paper that conclusively states that poor fascia equals cellulite. And the cellulite experts are studying everything but fascia! It's a societal oxymoron that's left us trapped like our bad fascia, with no answers. Of course, until now. *This* fascia guru is connecting the dots to study and cure cellulite! In fact, by the release of *The Cellulite Myth* our research will be concluded, and you should be able to read about it in the headlines.**#Goals**

5
Cellulite "Treatments"

*"No problem can be solved from
the same level of consciousness
that created it."*

–Albert Einstein

Now that you have a newfound understanding about your body and the fascia system, let's take a look at the current approach to cellulite. As you have already learned, there are many factors that impact the condition of the fascia. Addressing these factors will provide some improvement, which is why you may have seen a level of results from something you've done previously. However, there are limitations if you are not directly restoring the fascia because—say it with me—cellulite is a fascia problem!

This chapter will specifically cover three categories of treatments currently available to consumers: invasive, noninvasive, and at-home treatments. I will be making some broad generalizations about these approaches and I know there are exceptions out there. (**#DontInboxMe**) I just want to paint a broad brushstroke so you have an idea why something may or may not have worked for you in the past. Or if you are considering any of these options, you can have an idea of what to expect or what you should be researching further. Although, by the time you are done with this book, it's doubtful you will be interested in any of these.

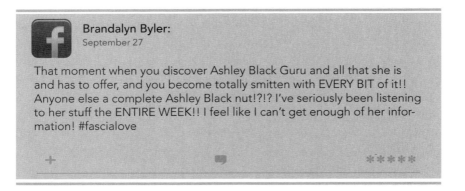

INVASIVE CELLULITE TREATMENTS

Invasive cellulite treatments are treatments that are done inside the body, also known as plastic surgery. For the most part, the plastic surgery industry knows that cellulite is a fascial condition. However, they tend to call fascia: membranes or connective tissue. As I wrote earlier, this is what WebMD® calls it. They know the root cause of cellulite is fat pushing up through the connective tissue, but their solution is invasive surgery. The surgical procedures focus on removing the fat by sucking it out in some way, or melting it with injections, or through fat lysing. They understand that the fascia needs to be addressed, but they are using completely different methodologies that I could never personally ever recommend. They have procedures to stretch the fascia internally, or pull the fascia bands, while sucking the fat out with a cannula. There are FDA-approved procedures to actually cut the bands of fascia (Interstructural/Ace® Bandage) with a laser or scalpel, which will release the tension and puckering. However, over the long term, this will cause more internal scarring, which is the fascia "rebelling" against the invasive treatment through recoiling. I would not recommend cutting fascia as a means to reduce cellulite. Fascia is an important system of the body and, from what we currently understand, the fascia would regrow ultimately causing more R.A.D. than before.

With all due respect to the industry, they are onto it. They understand the fascia needs to be loose and the fat needs to be gone. They just do it through cutting and surgery which is the nature of the field.

By the way, if you have had any of these procedures, whether you experienced a good or bad outcome, it doesn't negate the need to restore your fascia. Regardless of what you've done in the past, everyone needs a healthy fascia system. (Of course, you probably already know that and that's why you are reading this book.) The methods in this book are noninvasive and people have reported better results than with invasive surgery. A great testimony is on the back of this book, and there are thousands upon thousands more in the FasciaBlaster® Facebook® group. (Thanks Karen!)

If you are in the group of people who have had really bad plastic surgery experiences, particularly liposuction where the cannula has poked through and altered the fascia, restoring the fascia will help you break down the resulting scar tissue. Remember, scar tissue and fascial adhesions are chemically the same. Stay tuned!

NONINVASIVE CELLULITE TREATMENTS

Noninvasive treatments include anything done topically or manually to the skin. I must say, the developers of some of the creams out there are definitely on to something. Not only do certain creams and oils have some effect, I created my own line of complementary products for this reason. There is solid science that shows how you can stimulate fat burning through the skin via thermogenesis. The way that creams approach cellulite is, again, by focusing on the fat problem instead of the fascia problem. You need to do both. There are creams that increase blood flow, and that is good for fascia, but it's not a comprehensive, restorative option. Using creams alone to address cellulite is like trying to get across the ocean on a dinghy when you can use a private jet! (My methods are the private jet!) Creams and oils are the cherry on top, not a total solution. There are great products out there and I suggest using them, especially mine! Yes, with a team of amazing scientists, I formulated what I call Blaster Oil™ and in clinical trials it has shown to "blast away" cellulite on it's own. I will tell you more about how to use it when we get into the protocols of this book or you can go to AshleyBlackGuru.com/TheCelluliteMyth for more information.

HEALTH AND FITNESS INDUSTRY

The approach from the health and fitness industry to get rid of cellulite is also considered a noninvasive approach. Their goal is to impact the fat through proper nutrition, which is good for fascia and one piece of the puzzle. Good nutrition affects the body at a cellular level. If you are doing something good nutritionally, then you are doing something good cellularly. If you are doing something good cellularly, you are doing something good for your fascia. However, that's not the whole story.

The industry also focuses on weight training and building muscle, which is also good for fascia because it flushes blood to the area, and blood is restorative. Also, if you are a smaller person with loose skin, building muscle will fill up the loose skin and improve the appearance of cellulite. On the other hand, if you are muscular enough and lean enough, you won't have the look of cellulite, but it doesn't mean your fascia is healthy. The fascia can still be tight and restrict muscle fiber access and blood flow, and you won't even know it. If you've ever had trouble building muscle, this is probably a huge reason why.

Here is what bothers me the most about the health and beauty industry, the appearance is the measuring stick of success. It's true— beauty is only skin deep. However, you shouldn't have to sacrifice your health to be beautiful, nor should you have to sacrifice your beauty to be healthy. With my methods you can have both!

I give the fitness industry props because nutrition and exercise will reduce fat and build muscle, which is all good for fascia, but again, it's not a total solution. I also want to mention that yoga, body work, acupuncture, foam rolling, stretching, massage, etc., are all positive things for the body and fascia system, *when done properly*. Most people are not doing them properly and damaging the fascia as a result. Having said that, even if you did all of these things and had a perfect diet and exercise regime, you still need some way to specifically address the fascia.

AT-HOME CELLULITE TREATMENTS

Okay, first of all, major props to my DIYers! If you do dry-brushing, essential oils, rolling pins, supplements, or anything on your own to address your cellulite, hats off to you! What I like about DIYers is that you are taking matters into your own hands, which is exactly what you need to do if you want to treat your fascia successfully. Getting rid of cellulite means addressing the fascia regularly and ideally, a little at a time every day. If you are a DIYer, you are already mentally geared up for it. **#GOYOU!**

Now, I just want to stop right here and take a minute to say if you are doing something that you believe is helping your body or affecting the appearance of your cellulite in some way, if you are a dry-brusher, supplement-poppin', essential oiler, mama's homemade cream, hot yoga doin' . . . or bioidentical hormones person, I'm not at all telling you to stop those things. Whatever is working for your body, do it. You are the only person who ultimately knows what your body needs. Addressing the fascia in the manner I am going to describe will enhance everything else you are doing to create homeostasis in your body. There are many pieces to your body puzzle so keep collecting them until you have a full picture. Consider this understanding of fascia to be the exclusive "peek behind the velvet ropes" that you may not have known previously.

The reason I like DIYers, and the reason I invented a DIY product that I will tell you about soon, is because the fascia needs regular attention. I have found that if you aren't the one giving your fascia that attention and you are relying on someone else, your "me time" (or fascia time) can easily be placed on the back burner due to financial constraints, time restraints, or simply convenience. If you can't give your body what it needs yourself, you get stuck in the hamster wheel of depending on other people for solutions. The DIY approach puts you in the driver's seat of your own body. The education I'm providing will show you the way.

Now, here's what I don't like about DIY cellulite methods: there are very few options out there that actually work so if you've been putting in maximum effort for minimal results, everything is about to change. Keep reading!

MY APPROACH—THE FASCIABLASTER®

As I have mentioned throughout this book, I have researched and worked in fascia for over 15 years. My clients were primarily pro athletes and I was searching for a way for them to be able to care for their bodies when I wasn't around. I was flying all over the country to treat my players but I obviously couldn't be everywhere at once, so I was looking for something to fill in the gaps. I couldn't find anything on the market that could do what I needed it to do and so, with my engineering background and my DIY personality, I began to design a tool myself. I wasn't trying to invent the cure for cellulite; I was trying to invent a fascia tool. I wasn't even looking to market it publicly; I just wanted a tool for my clients to use as a supplement to my treatments. At the time, I was running fascia centers and looking for a way to duplicate my techniques with an at-home method, giving my clients more control.

When I got the prototypes of my tool I quickly put them in the hands of my clients. It was so unique and effective that soon their pro teammates wanted it, and then family and friends wanted it. One day, one of my NFL® players gave it to his girlfriend, who was *highly* fit but who had a little cellulite. When I say highly fit, I mean highly fit and lean. She called me shortly after she started using it and said, "Hey, Ashley, that fascia stick that you gave us got rid of my cellulite."

And that is when the heavens opened and the angels saaaaang, ya'll! The roof opened up and a light shone down upon me as I sat there stunned and processing the implications of that discovery. To say it was a larger than life moment is an **#UNDERSTATEMENT**!! Here I was, a pioneer in the fascia movement, and it never even dawned on me that fascia and cellulite are the same thing, even though I had cellulite my whole life and mine was strangely disappearing. I just hadn't thought of it before. OF COURSE cellulite was fascia and I had the cure for it— the FasciaBlaster®. My team and I set out to do official testing and saw amazing results in our double blind studies with original test subjects. We see astounding results every day with tens of thousands of users, and the big budget test results should be released about the same time as this book. Visit AshleyBlackGuru.com/TheCelluliteMyth for more information on testing and results!

With the FasciaBlaster® and the knowledge in this book that is based on real science, anyone can have a profound impact on their bodies as real as any plastic surgery, but safer and more affordable. **#WelcomeToWonderland**

What makes the FasciaBlaster® unique is the way it breaks up the fascial adhesions, loosens recoil, and smooths distortions, which in turn releases restrictions. As a global statement, any device that rolls or smashes over the fascia will not restore the fascia. Think of R.A.D. issues in the fascia like a tangle in your hair. Rolling over the tangle will not de-tangle it. A hair tangle needs to be raked through, layer by layer, until the hair is tangle-free. Restoring the fascia is somewhat like getting a hair tangle out. To get rid of cellulite you need to be able to break up the fascia and smooth it out, in essence, realigning and restructuring it like a hair tangle. Blasting through fascia makes a night-and-day difference compared to rolling over it!

There are other tools out there marketed for cellulite, but they are based on the work of people who are studying primarily the surface fascia. There are some methods that anchor the skin and scrape or stretch it, which does impact the fascia on some level, but it doesn't address or break through the R.A.D. deeper in the body, which also cause cellulite.

Also, more often, we are hearing about home cupping. This is a treatment that takes small suction cups and attaches them to the body. Cupping is good in that it brings blood to the surface. It's also creating a tiny stretch in the fascia, but again, it's not breaking through adhesions, ending recoil, or smoothing distortions. Cupping also cannot restructure the fascia. I'm not saying don't do it; I'm saying know what to expect when you do it. It's good for blood flow, but limited in regard to fascial restoration.

There is also a little massage tool that exists with legs that may appear to work like the FasciaBlaster® but again, it will have limited results. When I designed the FasciaBlaster®, I considered the size of the claws, length of the bar, and overall weight so a person could use it on every part of their body and have the right leverage for maximum impact. There was so much attention given to every detail of the design that my team and I never really cared about what it looked like. It was the feeling and functionality that mattered and at the end of the day it works 1,000 times better than I ever thought it would! It's fixing conditions I've never even heard of and I couldn't be happier about all of it!

As I was writing this book and discussing it with my agents and publishers, we wanted to make sure this was not about selling a product. We considered writing protocols to scrub your fascia with your knuckles so people would not feel obligated to purchase a FasciaBlaster®. But ethically, I couldn't do it. There is truly nothing like the FasciaBlaster® and anyone who has one will tell you! If this were a book about juicing, I wouldn't expect you to make kale juice by squeezing the heck out of the leaves manually; I would tell you to get the proper tool for the job—a good quality juicer. Same idea. The FasciaBlaster® is the "juicer" for addressing cellulite. Right tool equals right result, or optimal . . . juice!

Since this book was written, we've introduced additional FasciaBlaster® tools and products. To move forward with the protocols in this book you do need at least the original FasciaBlaster® tool. I would imagine the majority of readers have one, but if this is all new to you, now would be a good time to order your FasciaBlaster® at AshleyBlackGuru.com/TheCelluliteMyth!

Julie Thrane Sheckman ▶ *FasciaBlasters*
March 20

Ashley Black and the FasciaBlaster® became a part of my life in August of 2015. When the ad for the FasciaBlaster® flashed across my Facebook® feed I thought "Why not? I've spent money on way crazier stuff!" The first 30 days I learned to use my FasciaBlaster® with noticeable differences in my body and cellulite. At 60 days I leaned out and lost 10lbs. 90 days in I knew that this tool was transforming my life and my body. Fast forward one year later:

Here are my results:

I lost 15lbs overall.

I lost over 2 inches off of each thigh and 1 inch off of the dreaded knee fat.

My thighs and saddlebags are transforming from Beyond Bound™ to perfectly smooth legs (I can pinch less than an inch!)

My Cesarean scar is gone and a facial scar is going.

My spider veins are going, going, almost gone.

The ringing in my ears is less.

My misshapen rib is straightening out.

My knees no longer pop.

Here's what I've found:

Confidence, energy, and a zest for life!

Flexibility I thought I'd never have.

A sense of being a part of something very special that has and is changing lives every day

New friends from coast to coast and around the world

I am 57 years old and forever grateful to Ashley Black!

#grandmaswhowearbikinis

✦   🗩   ✶✶✶✶✶

Tiffany Marie Herrington
December 4

So I joined this group when there were only 5,000 or so ladies here. I searched high and low for a "cheaper" alternative to the FasciaBlaster®. I finally ordered the real thing and did my first blast last night. I can whole heartedly say there is no comparison! The FasciaBlaster® produces such a different sensation. Save your money and get the real deal!

6

Analyzing Your Fascia

*"What lies behind us and
what lies before us are tiny
matters compared to what lies
within us."*

−Ralph Waldo Emerson

Now as you know, cellulite aka crappy fascia, is an equal opportunity hater, and it hates on people in four main ways. These are what I call the four types of cellulite. Most people have one or more of these four types in various places in their bodies. You can have one type on your leg, and another type on your arm. Knowing your cellulite type is one step in the process of analyzing your overall fascia so you can restore it, get rid of cellulite, and be the healthiest version of yourself possible!

HAIL DAMAGE

The first type of cellulite is Hail Damage. It appears lumpy and bumpy and is probably the most common presentation, but it's also the fastest and easiest to change. If you think about it, hail damage on your car happens at the very surface of your vehicle, and Hail Damage on your body happens in the most peripheral or surface layer of the fascia, about a quarter of an inch below the surface in the Structural/Ace® Bandage fascia.

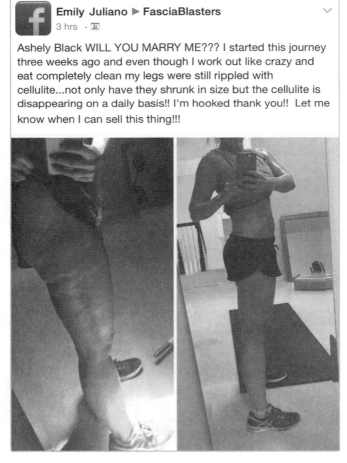

The easiest way to visualize this type of fascia is to think of the "skin" of a sausage casing. (Sorry if you're not a meat eater, but the visual really works.) To make sausage, meat is ground up and a casing is formed. The shape of the sausage isn't based on the amount of fat in the meat. The shape of the sausage is determined by the size and shape of its casing! If the casing is dimpled like a golf ball, then your final sausage shape will be a dimpled golf ball. If the casing is shaped like Mickey Mouse®, the sausage will look like Mickey Mouse®. Now, imagine if that casing were a pair of fishnet panty hose. What do you think that sausage will look like? Yep! The meat will poke through the tiny holes and look bumpy, dented, and dimpled. Sound familiar . . . maybe like someone's legs you know? When the fascia is unhealthy, it distorts and tightens and pulls and fat bulges through the thin areas just like ground meat stuffed in fishnet panty hose. (For you vegetarians, visualize a giant marshmallow being stuffed through a chain link fence!)

This image represents fascia that causes the look of Hail Damage cellulite.
It is a marshmallow being pushed through a grid.

This image represents smooth fascia and cellulite free skin.
It is a marshmallow pushing against plastic wrap.

When you address the fascia, the goal is to restore it to its optimal, smooth state by changing the shape of the outer casing. FYI, if you are thin and have loose skin, loose skin exacerbates the look of Hail Damage. When you add muscle it plumps the skin and reduces the appearance of Hail Damage. Fair warning, if you have this type of cellulite, you are going to bruise like crazy as it is restored and opens. Don't freak out; it typically doesn't hurt as badly as it looks.

CAR WRECK

The second type of cellulite is Car Wreck, and the name says it all! This is where you have dents, lines, or divots, which indicate R.A.D. (restrictions, adhesions, and distortions) in the deeper layers of fascia. Straight up, your skin just looks like a car wreck. Sorry to be blunt, but it is what it is and the truth will set your fascia free! I mentioned this in chapter two as well, but the reason you have dents and divots is because the deeper Interstructural/Cotton Candy fascia has R.A.D. and is pulling down on the Structural/Ace® Bandage fascia like tufting. The important thing to understand about restoring Car Wreck is that you have to first restore the surface layers of Structural/Ace® Bandage fascia before you can dig down and restore the fascia pulling from deeper within.

> When you address the fascia, the goal is to restore it to its optimal, smooth state by changing the shape of the outer casing.

Anyone can have Car Wreck cellulite. It doesn't matter if you are as skinny as models or fuller figured and voluptuous. Again, you can have mixed types of cellulite all over your body. You may have Hail Damage on the front of your legs and Car Wreck on the back.

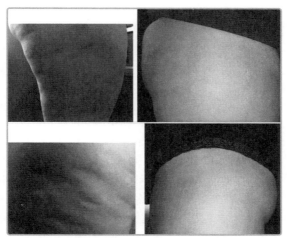

The images on the left are "Car Wreck" cellulite and the images on the right are the results of the FasciaBlaster®.

GUMMY BEAR

Gummy Bear is our third type of cellulite. While gummy bears are cute, when it's your legs we're talking about, you don't want to be gummy. The appearance of Gummy Bear cellulite is the result of atrophied muscles, fat, and fluid clumped together. The tissue just feels gummy, without definition. That's because the muscle in that area is so restricted by tight fascia, due to, say it with me, R.A.D.—restrictions, adhesions, and distortions—that the nerves aren't able to fire a strong enough signal to make the muscle fibers fully work. It's like trying to get a cellphone signal inside a concrete room. The signal is just blocked. Tight fascia impedes, or blocks, your nerve signaling. Also, the blood is not able to flow sufficiently to that area in order to build and strengthen the muscle. To make matters worse, the tight fascia is also trapping fluid (inflammation) in the area and may be causing swelling. This makes you look bigger than you are on top of any fat that is there hanging around doing what fat does.

If you are restoring Gummy Bear cellulite (fascia), you may go through a brief period where the "jiggle" of the skin becomes more pronounced and may look like it's getting worse, but it's actually getting better. You are breaking up the larger chunks of fascia and it is becoming more pliable so that you can have access to the muscle. Remember, the underlying muscle is soft and atrophied. By opening the fascia, the nerve signal can now fire more efficiently and the blood can flow more optimally to the area so that you can properly build muscle. Muscle tone will shape the gummy areas as you restore the fascia so that you don't end up with droopy skin but instead, have the beautiful shape that comes with having healthy fascia!

In the coming protocols you will learn how to open the tissue so you can activate the muscle. To "activate" a muscle simply means you are working the muscle fibers by using them. When you activate, which can be as simple as an isometric contraction or weight bearing movements, you are building the neural connections between the brain and the muscle and supporting overall muscle strength and growth.

The bottom line is that correcting Gummy Bear is all about opening the tight fascia so you can have better access to muscles so your body can function more efficiently, and you can build muscle definition— sometimes for the first time ever.

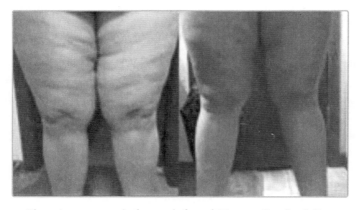

These images are a before and after of "Gummy Bear" cellulite.

BEYOND BOUND™

Finally, we have Beyond Bound™. Beyond Bound™ is the worst of the worst type of cellulite—but take heart, because it's still fixable! Fixing it will just take a little longer because the R.A.D. is in all of the layers of the fascia. When you are Beyond Bound™, you are flying off the phases of the Fascia Freak-out Chart with symptoms galore tied to your fascia condition. With Beyond Bound™, the fascia is so tight that you can't even pinch or pull up the skin because you have solid bound chunks of fascia underneath it. A person who is Beyond Bound™ through most of their structure feels sluggish a lot of the time, with poor circulation. It's hard to perform an exercise, leg cramps are common as well as a host of other physical conditions tied to the poor quality of the fascia. But like I said, have hope—it's fixable!

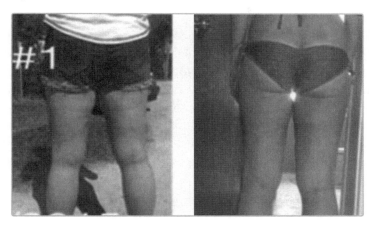

These images show the before and after of "Beyond Bound™" cellulite. Although the dimples are less than other types, the "Beyond Bound™" is the most severe.

Remember, you didn't get this way overnight and so it will take some time to undo it. This is a process. Time is passing anyway so you might as well turn things around and know you are making progress instead of further deteriorating. Beyond cellulite, beyond beauty, beyond your appearance, I want you to be as healthy as you can be so you can enjoy life to the fullest. Too many people are exhausted at the end of the day, in pain, popping pills, and "vegging out" in front of media, because they have no energy or motivation. This isn't the way we are supposed to live! When you are pain-free and feeling healthy, so many possibilities will open before you! I want you to see yourself living as healthy as you can be, which means having really good fascia! When your fascia is healthy, the rest of your body systems have a better chance of being healthy, too! Healthy fascia means your body functions optimally, blood flow and nerve activity are free, and your skin is smooth and supple. It's all within reach! The question is, how hard do you want to work for it? Achieving your body goals takes time and commitment—two things every person can give. **#CanIGetanAmen**

THE FIVE Ps OF ANALYZING YOUR FASCIA

Now that you know a little about the cellulite types, let's get ready to fully analyze your fascia. There are five self-tests for this, and we were just cute enough to figure out how to make them all start with the letter "P"—Perception, Pinch, Poke, Posture, Position. Not only should you go through these five steps when preparing to address your cellulite, but you should go through these five steps if you have any pain, restrictions, or issues in your body whatsoever. And, you should do these periodically as you are employing the protocols so you can monitor progress beyond what you can see. Do this *before* you go running to a doctor or start popping pills, before you go to a trainer, therapist, or any other type of body worker. Girlfriend, as my Mama always says, "Start where your butt is," simply meaning, "What are you doing right now?" Don't go taking your butt to everyone else on the planet and expect them to know what's going on. If you're having a problem, ask yourself these questions: Am I hydrated? Am I stressed? Am I reacting to something I ate? And then go through the following steps to analyze your fascia. DISCLAIMER: I'm not saying you shouldn't go to doctors. (Since 98 percent of you get that.) This is for the 2 percent who are getting ready to inbox me. Stop right there! All I'm saying is that you should educate yourself about your own body first. Then you can have an educated conversation with your doctor, if you need to go that route. **#YouBeTheExpert**

> It's time to give yourself permission to like yourself.

Moving right along, this process of analyzing is a very simple, common-sense approach to your own body. In fact, in the medical community they call this process H.O.P.S., which stands for History, Observation, Palpation, and Special Tests. That's what we're doing with your fascia. Basically, the 5 Ps are about finding "where your butt is" and looking at in a scientific way. Instead of running down the crazy trail of, "Should I do this, that, or the other?" Stop and look at where you are right now so you can make some educated decisions.

Also, I want to point out that you will use the first three steps to determine your cellulite type. The perception test is, of course, about what you see, and cellulite is, of course, what you see as a symptom of unhealthy fascia. The pinch and poke tests also tell you about your cellulite and where it's coming from; however, cellulite is just one piece of the fascia puzzle. If you want to get rid of cellulite, you have to address the entire fascia system as whole, so we need to do some more investigating, which is where the posture and position tests come in. Remember, when the fascia is healthy, the symptom of cellulite will disappear! (For the 489th time . . .) **#CelluliteIsFascia**™

No Crazy Head Allowed

Before we really get going here, I just want to say, every woman is a little crazy about the way she sees herself in the mirror. Society is hard on us and so many women are looking for the affirmation and encouragement that perhaps they didn't get foundationally growing up. (Yes, I've been to therapy, too. **#Eyeroll**) Sadly, instead of celebrating inner strength and beauty, women have become their own worst critics, tearing down their own bodies. When was the last time you looked in the mirror and said, "Gee, I really like my _____?" Anything? *Evah?* It's time to give yourself permission to like yourself . . . and if that works out, try loving yourself! All of you!

Remember this: when you are looking at your body and analyzing the condition of your tissue, *no crazy head allowed.* I'm asking you to take off the lenses of criticism and put on the lenses of a detective. This isn't about pointing out all of your flaws; this is about finding the root of your issues and employing solutions. Think of yourself as a remodel project. If you are remodeling a house, you don't get too upset about the condition of the floors because you already know you are going to replace them. No big deal if there's a scratch or dent in the hall, because demo day is on the way! If you are looking at the walls, and you don't like the paint color, there's no reason to burst into tears; paint can be changed. That's the attitude you should have when looking at your body, all the way through the process. When you remodel a house, sometimes it gets messier before it's finished, but you are still making progress!

You may not have ever realized your own power to change and sculpt your body, but trust me. The notion that says we don't have control over our genetics or the way our bodies are shaped, or cellulite, is a MYTH! If you've been on my FasciaBlasters™ Facebook® group, you know that you can change any part of your body that you want, and tens of thousands of women have posted pictures proving it. You are on a path to make changes to yourself so don't freak out by what you see now; instead, be empowered by what you can do about what you see. When you start learning how truly amazing your body is, you really won't be concerned about little imperfections. Knowledge is power and power brings confidence! So take the emotion and judgment out and just say "no" to crazy! (**#GotMyPreachOn**) Are you ready to move forward? Let's do this! **#NoCrazyAllowed**

Melissa Vaughn Thayer
March 18

I have had wonderful health benefits from using the FasciaBlaster® and really wish I'd taken more before pictures. I was diagnosed with MS in 2002 and was on 12 different medications. Since I started blasting a little over three months ago I haven't had any MS medication at all. No steroids, no Avonex injections, nothing. Matter of fact, I went to the optometrist and he had to send for my old records because there was absolutely no sign of MS in my eyes at all! I eat clean, I blast, and I work out 5 times a week. I feel better than I ever have and I don't even take my a ADD meds anymore. I have been on Adderall for 15 years and now I take absolutely nothing except for vitamins and supplements! I have my life back and I feel great!

＋　　　　　　　💬　　　　　　★★★★★

1. PERCEPTION TEST

So right now, go take off all of your clothes and stand in front of the mirror and look at yourself! (Unless you're not home. Please wait until you get home. And preferably with no one else around.) Take a note pad with you and for Heaven's sake, close the blinds. Also, grab your phone so you can take some full body pictures. Yes, you can do this. Try to get in an interior room so you can control the lighting. Pictures are the best way to see changes on this journey. If you can have someone else take them that's even better—and you may want to take a set wearing a swimsuit so you can share your results, receive coaching, and inspire someone else! Take full body pics, front, back, sides, and "parts" you want to improve. You can inbox them to my team one of two ways: 1. Inbox the Ashley Black Guru Facebook® Page or 2. Post in the private FasciaBlaster® Facebook® group for personal feedback and encouragement.... please no nudes! Trust me, if you don't take pictures now, you'll wish you had! Suck it up and capture the moment!

Okay, are you standing nude, or almost nude in front of the mirror? Stop for a moment and just look at your amazing body. What do you see? What has your body done for you over the years? What parts of it do you like? What parts do you want to change? Look at the front . . . back . . . sides . . . under your arms . . . at your knees . . . your elbows—everything. Do you have your detective lenses on? You may feel like you're at a crime scene, but don't you dare slip into crazy on me. Even though "CSI" is going on, take heart; you will solve this mystery.

Do you see some Hail Damage or maybe some lumps and dents? Are there areas that are soft and gummy? Or maybe Beyond Bound™ chunks that are just solid and tight. Don't worry if you're not exactly sure of what you are looking at; just start taking it all in. Maybe you've never looked at yourself completely naked before. Maybe, because of past violations, it's hard to be this vulnerable, even with yourself. We've had reports of emotions being stirred in women, but in the process, healing comes. Don't be surprised if you start feeling emotional unexpectedly. It's okay. Embrace the process. Sista, you got this! This is the first step on your journey of transformation. Change isn't easy, but like that butterfly fighting to come out of its cocoon, it's time for your coming out party. **#Yaaaasssssss!**

While you're observing your body, did you lift your arms and find "bat wings," aka, underarm flab? Well, hop down this little rabbit trail with me for a minute on that topic. When you hold up an arm and see the skin and tissue hanging down, you might automatically think you have a flabby triceps. However, if you were to turn your arm upside down, all the extra skin would fall in the direction of the biceps. Flabby arms are a gravity and an extra skin problem, not just a triceps problem. If you need to fill up saggy skin in your arms, work on both biceps and triceps. But know this—saggy skin and cellulite are two completely different things; however, the FasciaBlaster® will address both!

Here are some points to consider during the Perception Test:

1. What looks good is not a true measure of what is actually good, or healthy. For example, Beyond Bound™ looks way smoother than Hail Damage, but it's way worse than any other type, because your fascia is on lockdown!

2. Depending on where you hold your fat, have muscle, or where you have loose skin, it can affect the way your cellulite/fascia looks.

3. You can have dysfunction in one area that is visible in a different area. If all you are doing is looking at your body, you will miss the root cause of some of your issues.

4. If you want to, go ahead and make some notes about what you see. You don't have to, but it may help you connect some dots later on and help you monitor your progress. Where do you see flabby skin? Where do you see dents on your legs? You might even want to note how you are feeling right now. Do you have pain anywhere in your body? What emotions do you have about this process? The internal transformation can be as great as the external process, so pay attention to everything that is going on inside of you. Again, don't stress about this or go into crazy mode. If you want to make notes, do it. If you don't, don't. You are in the driver's seat, baby, and it's your car! **#FasciaFreedom**

Crystal Ferren
March 8

One day as I was scrolling through Facebook®, I came across an ad, for cellulite reduction, with this beautiful lady holding this weird looking stick thing! As I thought to myself, "Here we go, another scam!" I clicked on the ad anyway and started reading some of the information about the tool. I then sent my pictures to Ashley. To my surprise, she got right back to me. So, I hesitantly ordered my FasciaBlaster® that same day. (It helped that there was a 60-day money back guarantee.) I didn't realize it at the time but that was the day my life completely changed for the better!

I bought the FasciaBlaster® to help with my cellulite not knowing all of the other benefits. Yes, my cellulite has reduced significantly and I have more confidence in myself than I have ever had in my life! You see, I was diagnosed with carpal tunnel about 12 years ago and the medical professionals suggested surgery. I refused it. It was a pain that I just dealt with day in and day out. It was hard for me to close my hand to make a fist or snap my fingers, or even use a knife so I could chop food while cooking, which is my passion. For me, it was difficult to do those everyday little things that most people take for granted.

Amazingly, I used the FasciaBlaster® twice on my arms and hands and the pain was completely gone! GONE!!! I had a bunion and blasted that and the pain was gone. I also had knee surgery when I was 16. As I got older, my knee started causing me problems and I couldn't work out as much, or even walk around a lot. I would hear this loud crackling sound when I bent my knee like Rice Krispies® and my knees would swell up a lot. I was just a mess. However, a few blasts to the legs and knees and I was a whole new person! I also used to get crazy migraines, and since blasting I haven't gotten any! I've also noticed that I don't get cold sores anymore and I can wear heels for a longer period of time. I use the FasciaBlaster® on my fiancé, mom, and son. They have all experienced pain reduction and love the benefits that it has to offer!

I just can't believe how much Ashley Black has changed my life! I had made peace with myself and thought I would have to live with cellulite and all the aches and pain I was having. Isn't that what we are taught, that as we get older things just start to fall apart? Well, not anymore, thanks to Ashley! I can't even describe how happy I am because the pain is gone and I feel so free! I AM A NEW WOMAN!!!

2. PINCH TEST

Selfie time is over! Now it's time to do the pinch test! The pinch test is one way to find tight, unhealthy fascia that we can't see visually. Pinching tells about the fascia directly under the spot where you are pinching. Where the fascia is healthy, the skin is not bound, and we should be able to pinch up the skin easily. If you can't pull up the skin you have crappy fascia in that area, so make a note of it.

To analyze your fascia with the pinch test, you can pinch anywhere on your body. Note how much skin is between your fingers when you pinch up the skin and try to close your fingers. Just FYI, tissue is not supposed to feel hard. Look for Beyond Bound™-ness, smushy, globs of fat, or even "pebbles." If you have Car Wreck cellulite and pull on it, you will likely feel a burning sensation. If it's Hail Damage, you can pull it up and roll it in your fingers and it feels like rice crispy treats. Again, just make a note of what you are experiencing.

Here's what your pinch will tell you:

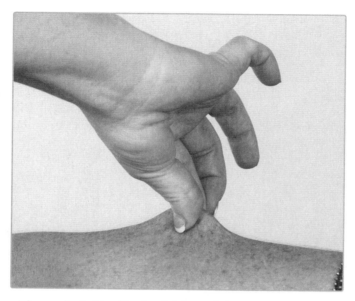

1. If you pinch ¼ of inch of skin and the skin pulls up, you have good fascia in that area.

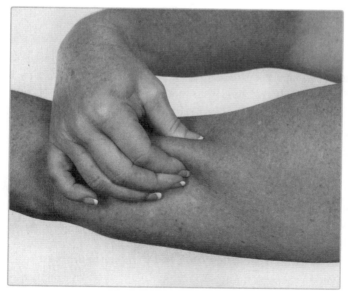

2. If you pinch ½ inch of skin, still good, but work needs to be done in that area to get it completely smooth.

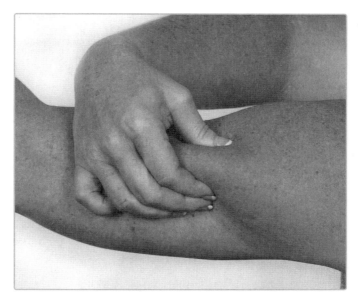

3. If you pinch a 1-inch chunk, you need to break this fascia up into smaller chunks and eventually stretch it out.

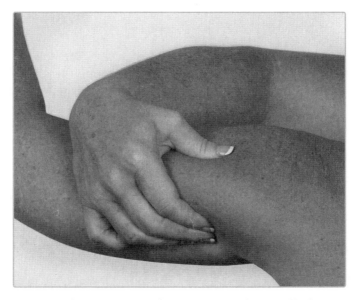

4. More than 1 inch, or if you can't pinch it at all, then it needs serious attention.

Amy Schwarz
February 11

I bought the fascia blaster hoping to reduce the appearance of cellulite on my butt and thighs. While waiting for it to arrive, I started researching and watching Ashley's videos. I was so surprised to learn that the pain I had been feeling in my IT bands for at least 20 years was my fascia. I had been to doctors and massage therapist and was resolved to the fact that "nothing" was wrong with me. I was also SHOCKED to learn that I couldn't pinch any skin anywhere on my thighs and that wasn't normal. I never thought that my pain and my cellulite were related. I am an RN, I've been to school, I'm medically educated, and this was ground breaking news to me.

For weeks I read and watched as much as I could find on the FasciaBlaster® and all things Ashley Black. I've been blasting since May 11th and am so excited to say not only is my cellulite going away. I no longer have pain in my IT bands!!!!! Also I can pinch the skin all over my thighs!!!! I'm learning new things about myself through this process and really I have just begun. I feel more in tune with my body, my work outs, my nutrition and my emotional well being. This is a new frontier and I am so excited about the journey I am on. Thank you, Ashley Black and your team for bringing this tool and your knowledge to the public!

✦ 　　　　　　　 🗩 　　　　　　　 ✳✳✳✳✳

3. POKE TEST

Now you are ready to move on to the poke test. The poke test helps you find problems that you can't see in the perception test or the pinch test. There are various points on the body for the poke test. There is no wrong place to do the poke test, but the image on the next page shows some suggestions. This is where it starts to get interesting, and you can actually see how tightness in one part of the body affects the appearance in another part of the body.

To do the poke test, put your finger on the specific point mentioned. This isn't an ordinary straight-on poke; you want to push with your finger away from the bone. As you apply pressure, you are looking for a change visually somewhere other than where you are poking. Make a note of where the change is. Do you see more cellulite? Did you see a line form? Page 114 gives an example.

If there is no change, note that, too. Where you see the change in appearance is where there is a fascial distortion, so make a note of that since you can't see it in the perception test.

Here are the points on your body where you are going to poke:

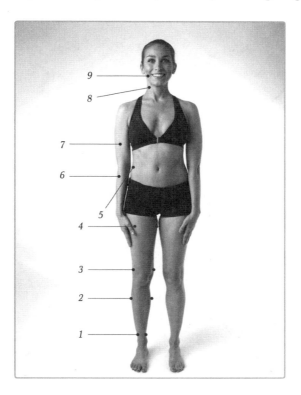

1.   Both sides of each ankle
2.   Both sides of each shin
3.   Both sides of each lower thigh
4.   Both sides of each upper thigh
5.   Both sides of each mid-oblique
6.   Both sides of each mid-forearm
7.   Both sides of each mid-upper arm
8.   Both sides of the neck
9.   Both sides of the jaw

Remember: There is no wrong way to do the poke test. Just poke and move the tissue while looking for changes that are consistent with R.A.D. in places you are NOT poking.

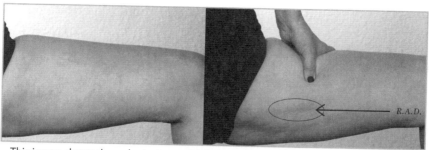

This image shows the poke test on a thigh. Notice how the fascia "puckers" in a spot that is not being poked.

Note: If there is zero movement when you press in, you are Beyond Bound™. Period. If it's super tender where you press, that's an area to work on too. You shouldn't be tender anywhere on your body by touching or pressing on it. That is unhealthy fascia!

Once you've done the first 3 Ps, you have all the info you need to determine your cellulite type and how to address it. You can skip to the next chapter, however, if you want to learn more about the condition of your fascia, continue on with me to the next test. Tests 1-3 tell you quite a bit about restrictions and adhesions, but tests 4-5 tell you about how your fascia may have distorted your structure. This will be helpful as you move through the protocols and make changes. Once again, feel free to post pictures to the private FasciaBlaster® Facebook® page for feedback and discussion, or just go take a look at what others have posted.

4. POSTURE TEST

Next, we are going to move into a more comprehensive analysis of the condition of your fascia by doing the posture test. If there's one thing that drives me absolutely crazy it's that so few people actually know what good posture is. (**#SoapBox**) It's like a contractor building a house without seeing the need to level the foundation. Anything built on that foundation is going to be unstable. Remember what it says in chapter 4 about structural integrity? If your foundational posture is jacked, all of your movement will be jacked and pain, injury, and cellulite will ensue! Let's get it straight and get our posture straight!

To perform the posture test, you need a clear space in front of a wall. Oh, and you can put your clothes back on if you haven't done that yet.

*Go to Resources on AshleyBlackGuru.com for a playlist.

1. Find neutral feet and knees: Stand with your feet and knees straight with even pressure in each foot with weight distributed between the second and third toe. Make sure feet are aiming straight like railroad tracks.

You can achieve this by leaning to the front, back, side to side, then end up in the center. The leaning helps some people with the concept of finding center.

Look down and make sure that both knee caps are facing forward as shown in the picture.

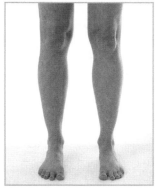

This image shows proper alignment and weight distribution of the feet and knees.

2. Find neutral pelvis: Make sure the pelvis is not rotated. You can achieve this by:

Rotating the left hip forward   Rotating the right hip forward   Ending in an un-rotated position. This is the correct position. Look down and make sure neither hip is more forward than the other.

3a. Find further neutral pelvis: Make sure the pelvis is not tipped to the front or the back. You can achieve this by:

  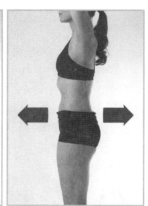

Anteriorly tipping the pelvis   Posteriorly tipping the pelvis   End in neutral position as shown. This is the correct position.

3b. You can check by placing 2 closed fingers behind the low back. Legs, upper back and head should be on the wall as shown in the image.

This images shows the 2 finger test to help find neutral pelvis.

4. Find neutral rib cage and mid back: Make sure the upper body is centered over the lower body. You can achieve this by:

Shifting the ribs to your right   Shifting the ribs to your left   End in neutral position as shown. This is the correct position. You can check this by lining up your nipples with your hip bone. (Yes, I said nipples!)

5. Find neutral head: Make sure the head is centered over the body. You can achieve this by:

Lining your chin up directly above the sternum, directly above the belly button.

Then, place your head on the wall and make sure you have a 4 closed finger space behind your neck. Make sure you also maintain the 2 closed finger space behind your low back while achieving the 4 closed finger space behind your neck. This is the "2 finger 4 finger test" that we discuss extensively in all our social media.

6. Turn on the correct postural muscles: Contract your lower abdominals and scapular depressors. You should feel your lower abdominals as if you are coughing or pooping (yes, now I said pooping) and you can engage your scapular depressors by sliding your hands directly down the sides of your legs.

All of THE POSTURE TEST is to be performed against a wall.

You have good posture if you can actually stand against the wall with your pelvis properly aligned, and your arms at your sides down by your legs, with everything straight and facing forward, *and* you have a two-finger space at the low back and a four-finger space at the neck. If this is you, then you have an actual, natural posture that you can base your other movements off of. If not, you have some work to do! (Most people do.) If you are standing in a dysfunctional posture, what makes you think you can walk correctly, or run correctly, or do any other movement correctly? The one thing you should be focused on is getting your body into alignment. As far as exercise goes, you should do simple cardio, like on an elliptical, or I've created a whole series of foundational movements that will help you clean up your biomechanics. Check out my website for more resources AshleyBlackGuru.com/TheCelluliteMyth.

The main takeaway from the posture test is to see where fascia may have been pulling, tugging, rotating, shortening, compressing, or distorting your structure. Poor structure means poor posture, which sets off R.A.D. and you know the rest of the story there!

5. POSITION TEST

We debated about whether or not to put this test in the book because it gives more of a broad brushstroke about what's happening in the fascia versus the targeted, detailed information you get from the pinch and poke tests. In this test, you will get into different positions and test your range of motion. These positions are stretching the long lines of Structural Ace® Bandage fascia. If you are limited in these positions, we won't know exactly where the distortions are based on this test alone, but we know it will be somewhere along that entire line of fascia. It also doesn't take into account a jammed joint that may be restricting your movement, or tight muscles or scar tissue, but it will give a quick picture of your Structural Ace® Bandage fascia. Your other tests may show you more about how tightness is affecting your movement patterns. The purpose of this test is to simply see how tight you are. It's helpful after you start working on your fascia to go back and check your progress by doing these positions again to see how your range of motion has increased. Ultimately, you want to feel the stretch along the complete line of fascia from one end of the stretch to the other.

It's not uncommon to start the protocols and eliminate tightness in one area, only to feel it in another area along the same line. That's because the original tightness was preventing you from feeling the tightness up the chain. Sort of like buttons on a super tight shirt—when one pops, the next one takes the tension. This is why you have to address the entire fascia system as a whole!

Structural Fascia Lines

The main fascia lines are depicted in the images on the following pages. These are visual representations of real anatomy, but because the muscles are also shown, the images don't convey the depth at which the lines of fascia run throughout the body. Just know these are not 100 percent anatomically correct, but will give you an idea about the placement and the direction of the lines of fascia within the body.

These lines in the Structural fascia tell you where you are actually jacked up. When you are looking for root fascia causes, two things that could be of issue:

1. Fascia line stretch—feels like a line of tension from intense to very subtle. Wherever you feel the most intensity in the stretch is where your fascial line is the tightest. If you feel even tension then you are equally tight or equally loose.

2. When bad, beware of impingement—this feels like a stab, pinch, or your body just won't get into the position. If you experience an impingement or impingements during any of the following positions, stop the test. You can try this test again after you have been FasciaBlasting® for a while. Impingements subside with continued use of the FasciaBlaster®.

1. Surface Back Line

Artistic representation

Position test to stress the line

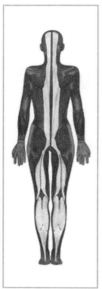

2. Surface Front Line

Artistic representation

Position test to stress the line

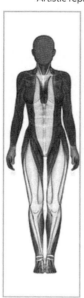

3. Deep Anterior Front Line

Artistic
representation

Position test to stress the line

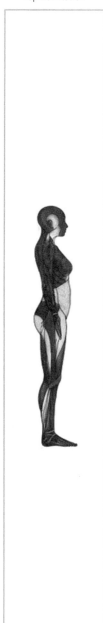

4. Arm line front

Artistic representation

Position test to stress the line

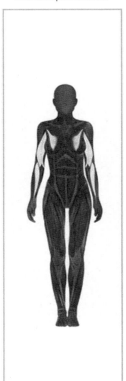

5. Arm line back

Artistic
representation

Position test to stress the line

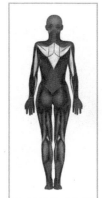

6. Functional line

Artistic representation

Position test to stress the line

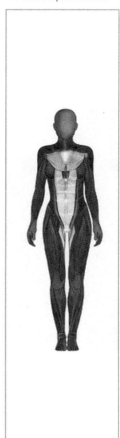

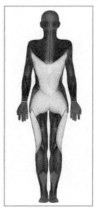

7. Lateral line

Artistic
representation

Position test to stress the line

8. Spiral line

Artistic representation

Position test to stress the line

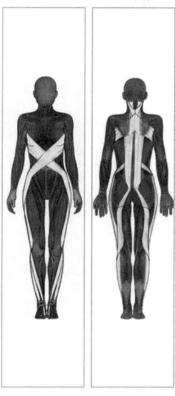

Right about now you may be feeling like you only asked for a drink of fascia and we put a fascia fire hose to your face. Hopefully though, you have a much greater understanding about the inner workings of your body. At the end of the day, people feel hopeless because they don't have knowledge. Now you know, and can have hope. So, next, let's look at what you can expect when you work on your fascia!

Laura Detering
October 30

HOPELESS . . . is one word that described me just 2.5 months ago. At 33 years of age, I had found myself hating my body more and more. Not just hating my physical appearance, but hating how my body was failing me. With 2 young daughters, I want to be active and I do not want to teach them to hate themselves, but, alas, I feel as though I may have been failing them, too. Hopeless failure. Two words that have haunted me. How did I get to my own pit of despair?

I have always been super active. I was a competitive figure skater for 10 years, played competitive volleyball for 7 years, dancer through the present (scratch that . . . I used to tear it up on the dance floor and now when I dance I pay for it for days), I did cardio for days, kickboxing, running, yoga, pilates, tae bo, P90X, etc. I can remember in junior high, compacting bones in my wrist while playing volleyball. I can recall falling hundreds of times in ice skating while trying to land my axel or my double salchow. I can remember even more distinctly my wrists during volleyball season swelling to the size of baseballs and bursitis flare ups in my elbows and continuing to play because my team needed me (more like I wouldn't let anything stop me). That was, until at the age of 18, my back would no longer let me enjoy life. I don't remember a single event causing the injury, but rather slowly I was getting worse. I couldn't sleep, stand for too long, sit for too long, without muscles spasming, spine twisting to the point where I was inches shorter on one side versus the other. Tons of tests later, they could not figure out why my lower spine was compressing. Ultrasounds identified small lumps on my low back as "injured tissue" that felt like a knife was stabbing me in the back at the slightest touch. Almost a year of physical therapy and chiropractic appointments got me much better, but from that point on, it was downhill and I never got back to where I was.

Fast forward to 31 years of age. At this time, I had a 4 year old, a rainbow baby, and an infant. I love my children beyond measure, but let's be real . . . this 3rd pregnancy did a number on me! The charlie horses in my hips from 3 months into the pregnancy were agony. A colicky baby that nursed every 1-1.5 hours 24/7 for SIXTEEN weeks took its toll on my upper body. My SI joint did not appreciate any motion besides walking or simple, weight training exercises . . . elliptical, seated bike, stairs? Forget it. At the slightest whim, I would sprain a finger . . . a finger! My neck, due to a car accident within a year of having my last baby, would not allow me to sleep in any position besides my left side (which further causes issues in my body). If I would lay on my right side, I would start spasming in the middle of the night as well as trigger substantial pain and if I lay on my back, and my fingers in both hands would go numb.

I started to become depressed. My saddle bags and deflated breasts weren't helping me to feel good about myself and my lack of physical activity was steering me deeper into the hole. My physical therapy from the car accident was not doing much besides offering me maybe 24-48 hours of some pain relief. I decided to not let it stop me and tried to help myself by starting to lift weights in the gym. If I couldn't do cardio because my SI joint wouldn't allow it, I would do something darn it!

Jump forward to earlier this year. By this point, I had lifted weights religiously for over a year. I was starting to feel better about myself, I had more energy and a much better sex drive. I was a better wife to my husband and better mommy to my kids and I felt like I was leading a better example for them on how to be healthy. However, I reached a point where results ceased to continue and, although I was lifting more weight, the scale was going up, up, and up. I simply looked "thick" and I was starting to hate my body again. My cellulite was just getting worse. I felt like my chest looked better, but at the same time, I all of the sudden had "pit tit" which may actually be because of muscle, but none the less, it looks like fat in tank tops. I had freaking crepe skin near my knees at 33 years old, people! Only 33 and I felt more like a 76 year-old woman who had never active a day in her life. I live in Florida and never show my legs and swimming or going to the beach was a daunting feat to me. How can I make memories with my kids if I am an anxious mom on the sidelines? I was desperate.

Desperate . . . . Now what?

I started researching cellulite removal. I knew lipo wasn't the answer, so I was trying to research holistic, laser, and other methods. I was also starting to battle guilt. Thoughts like, "We don't have money for these things; why do I care so much about my outward appearance when my hubby and kids love me as I am? Is it wrong for me, biblically, to worry so much about my outward appearance? Others have so many more terrible things going on in their life . . . what right do I have to complain?" These negative thoughts ran through my head all the time. But I just wanted to feel good again physically.

One day while checking in on Facebook®, due to Facebook's creepy stalking of my Google® searches, Ashley Black showed up in my feed talking about cellulite. I ignored it. Hmph. All my research says there is no cure so I will just need to accept that this is me and learn to be happy. After all, I do have so much to be thankful for.

About a week of seeing Ashley every day on my newsfeed, my intrigue got the best of me. Little did I know that my life would never be the same.

Today . . . BLAST ME!

I watched all of Ashley's videos. I checked her websites. I invited myself to the Fasciablasters group on Facebook® to see what all the women were saying. After talking with my husband, he told me to go for it! What was there to lose? It costs just about the same as one deep tissue massage. 60 day FREE trial, many women saying how it was helping them both with physical appearance, but also pain and emotional relief and a promise by Ashley and her team to help with individual coaching.

I am still considered a newbie being less than 3 months into the blasting phenomenon, but I am in it for life! Here are all the ways it has helped me thus far:

-It is smoothing out my body.

-My skin is looking better.

-I am showing more definition in my muscles.

-I am SHAPING my body little by little by blasting it using techniques taught by Ashley.

-My jammed hip is released.

-The compression in my low spine? Gone!

-Painful, injured tissue bumps in low back? All but 1 are gone and that 1 remaining is diminished

-SI joint flare ups? I blast it can and walk that same day . . . . It used to take me 3-4 days to be mobile again

-My posture has significantly improved

-My neck pain it significantly improving

-Better mobility and muscle activation!

-My butt is slowing shaping into a beautiful heart shape

-Plantar fasciitis flare ups are less often and less painful

-Choked muscles are coming to life (literally can feel it!).

-Ashley, her team, and this amazing blasting sisterhood are teaching me the importance of investing in myself and the proper ways to do so.

MOST IMPORTANTLY

-HOPE! I am no longer hopeless and the feeling of failure is fleeing. I am regaining confidence! You can see and check out my progress in the Facebook® blasters site :) I literally just showed my butt cheeks to 50,000+ women (Who does that?!?) I am showing my before and after pics to friends and describing my pain relief. I even blasted a friend who is a chiropractor and it fixed her hip that she's had problems with in one 5 min blasting session!! She will be offering blasting sessions with adjustments now. I have worn shorts and swimsuits this summer without stressing out about people seeing me. I am swimming with my kids and enjoying it! Most days I am pain free and when it does return, it's less and I can blast it away. Even though I have #goals that are yet to be achieved, I am committed to do the work because 1). I am worth it 2). My family deserves the best of me and 3). I am seeing and feeling the results!

If you have ever felt like me . . . like a failure, hopeless, fearful of what your body will feel like 20 years from now, guilt ridden . . . take it from me, pretty lady. You have NOTHING to lose, but everything to gain. Cheers to you and the first 60 days of the rest of your life!

7

Possibilities and Limitations

"Know your limitations, and then defy them."

—unknown

Before we get into the specific protocols, I want to help set your expectations of what is and isn't possible through this process of restoring your fascia with the FasciaBlaster®. The rabbit hole is pretty deep and vast as you are probably already discovering, but I want to make a few things clear. First of all, no one has all the answers, but the ones we have are pretty good. The FasciaBlaster® is so new that studies on all of this are still underway. We have a sample of nearly 100,000 women who share their experiences daily, so between their testimonies and my personal experience with my clients and staff, this is what we have seen and what we believe will happen for you.

POSSIBILITIES WITH THE FASCIABLASTER®

1. Restore fascia: The FasciaBlaster® is designed to palpate the connective tissue, which makes it more pliable. The FasciaBlaster® loosens R.A.D. (restrictions, adhesions, and distortions).

2. Smooth the skin: When the fascia is smooth underneath the skin, the appearance of the skin is also smooth. Additionally, healthy fascia promotes optimal blood and nerve activity and muscle access, which all contribute to smooth skin. Collagen production is enhanced and toned muscles fill up and tighten the look of loose skin.

3. Lyse the fat: Fat lysing is a process that simply breaks down fat cells. In the medical community this is done chemically and surgically; with the FasciaBlaster®, this is done manually when the muscle is contracted underneath the area you are blasting. This will be expounded on in another book. There are some limitations, which are discussed in the next section.

4. Increase blood flow and volume: When the FasciaBlaster® restores the fascia tissue, not only is blood flow in the current arteries and blood vessels improved, but the stimulation places a demand for the body to create new blood vessels and increase blood throughout the body. Blood is the life source of everything in the body.

5. Improve nerve activity: When the fascia is healthy, nerves that were previously choked by the tight fascia are freed to operate optimally; pain and numbness are relieved, and muscles fire more effectively.

6. Increase muscle access: Tight fascia restricts muscle fiber access. A person may be using a portion of the muscle or none at all. When the fascia is healthy, muscle fibers are freed and the improved nerve activity and blood flow activate their use. In essence, more "pump."

7. Change your bony structure: Bones are rigid and have no mechanism to move on their own. A hip or a pelvis cannot rotate itself. A joint cannot jam itself. It's actually the soft tissue, including the fascia, that rearranges where your bones are. This type of distortion is so compelling that it can compress individual vertebral segments on top of one another. This can also cause one

shoulder to be higher than the other. If you've been told you have a flat neck or a pelvis rotated, it's not the bone's fault. The bones are in the position in which the muscles and fascia put them in. Because the fascia is everywhere we certainly need to consider that the fascia could be a root cause. Loosening the fascia allows the bones to realign.

8. <u>Improve stretch marks</u>: When the FasciaBlaster® improves blood supply, skin healing can take place in ways never before imagined. The anecdotal research is undeniable, particularly when used with my oils and creams.

9. <u>Breakdown scar tissue</u>: At the end of the day, there is no chemical difference between scar tissue and fascial adhesions. Dramatic improvement can be made on all types of scarring.

10. <u>Reduce vein appearance</u>: Another result of improved blood flow is the reduction in the appearance of varicose veins and spider veins. Please see the next section on limitations.

11. <u>Reverse aging</u>: Another by-product of stimulating blood flow and smoothing the skin is that wrinkles, sagging, and age signs diminish.

12. <u>Improve mental capacity</u>: With improved blood flow comes improved mental capacity and mental clarity. Migraines and stress are reduced and overall mental well-being is vastly improved. You are getting more oxygen to the brain and decreased muscle tension and decreased fascial recoil on your brain. If you have ever had a headache so bad that you felt like your whole brain was in a vise grip, in all likelihood that was fascial recoil. Loosening the fascia will help.

13. <u>Improve neurology</u>: Healthy fascia promotes a better brain to body, body to brain connection. The FasciaBlaster® can reduce pain associated with neurological disorders. This will be further expounded on in a future book, just know that the science and possibility is there.

14. <u>Stimulates detoxification</u>: When cells are stimulated, the detoxifiers are stimulated. The skin is the largest detox organ of the body and when the fascia is restored, it's like unclogging a drain—the gunk is coming out. Also, the lymphatic system is stimulated by blasting.

15. Improve organ function: Once again, when the fascia is restored and not restricting the organs, blood flow, and nerve activity, the cells of the organs are healthier and everything functions healthier. We receive many reports of better elimination, hormone balance, and feeling of overall well-being.

16. Improve emotional well-being: There is scientific evidence about how the body houses memory and emotion. We have many confirming reports about how emotional awareness is increased with blasting.

17. Improve flexibility: When everything is loose, everything is loose! Range of motion increases and you just feel better overall.

18. Improve sports performance: This almost goes without saying, but when the fascia tissue is healthy and muscle access improved, all the supportive processes to sports performance—body symmetry, mobility, neurology, and so forth—are improved.

19. Improve libido: Do I really need to say anything here? Blood flow and nerve activity are key players.

20. Inhibit pain: Open fascia promotes tissue healing for all the same reasons described—increased blood flow, removal of toxins, better mobility, proper muscle use, and so on. The FasciaBlaster® also promotes nervous system desensitization, which is a fancy way of saying that it calms the nervous system and reduces the pain signal.

21. Improve overall well-being: There is something so incredibly powerful about taking back control of your body when you've been led to believe that so many things were impossible. When a person has control and begins to focus on the positive changes she sees, her self-perception changes. A new self-perception leads to self-love, and a reservoir from which a joyful, connected, fulfilled life extends. We were created to enjoy this life and when the body is optimal it supports living life optimally to the fullest!

While the above list is completely off-the-charts mind-boggling, and not even fully inclusive, there are some limitations I want to address as well. Restoring fascia is the key to the castle, but you still have to walk yourself into the castle if you want to enjoy it. There are some things the FasciaBlaster® can do, and there are some things *you* have to do to get the full result you are looking for—and some things we just haven't figured out yet. I'll go back to the house remodeling analogy. The FasciaBlaster® will do the demo and get rid of all the old, broken, dated, dysfunctional mess and prepare your house for the new and improved! But YOU have to do the heavy lifting on the reconstruction side, and sometimes, due to building codes, some things that you want to do just aren't possible. But the remodeled house will still look a whole lot better than the old one!

Kelley Zuniga
October 18

I've been using the FasciaBlaster® since March 2015. I'm a Personal Trainer and I've always had a tiny bit of cellulite no matter how lean I was. I initially bought the FasciaBlaster® to help cellulite and I didn't realize that it could help with my pain too! It has literally renewed by body!!

SO FAR the FasciaBlaster®:

Fixed my AC separation.
Fixed my ankle compression.
Fixed my tailbone pain.
Decreased flare ups from my chronic pelvic pain/pelvic congestion syndrome/overly tight pelvic floor muscles.
Stops my pelvic pain within 3 mins externally when it does flare up.
Gave me full access to my muscles..especially my hamstrings! I've Never had good hamstrings and can finally feel them like I should!
Improved my workouts tremendously!
Given me harder muscle contraction!
Ab sculpting and separation with no extra dieting or crazy ab workouts!!
Completely fixed my overall circulation, my legs no longer get frozen, my arms no longer tingle and I finally get red when warming up and blasting which means I have better blood flow.
Helped correct my naturally inward facing knees.
Took away my knee pain.
Fixed my hip flexor tightness
Helped increase my hand strength.
Took away my early on carpel tunnel symptoms.
Completely fixed my locked/frozen fingers too!
Took away my wrist pain . . . I can do push ups again!!!
Completely increased my flexibility, now I do yoga easily
Increased my ROM.
Gave me a boob lift!!

Overall, my quality of life is radically improved! I'm 31 and before, I felt like I was 80, very stiff. My body moves so fluidly now. I owe it all to Ashley and her team!

LIMITATIONS OF THE FASCIABLASTER®

1. <u>It cannot build muscle independently</u>: In order to build muscle, you have to send both a nerve signal and blood flow to the muscle by using it. The FasciaBlaster® will give you more access to muscle fiber and improve nerve activity, but you have to activate the muscle and actually start firing it or else nothing is going to happen. If you want to look like a fitness model with six-pack abs and rock-hard glutes, you have to work out like a fitness model with a six-pack and rock-hard glutes. The FasciaBlaster® will enhance your program and can speed your results, but you still need to have a program. The FasciaBlaster® is the gas, but you still need a car. Be realistic. If you want to see definition, you have to have muscle underneath.

2. <u>It cannot make you completely skinny or lean</u>: I know, I just got done telling you the FasciaBlaster® can help break down fat cells! And it can. However, it can only affect the fat cells it can reach. Fat is all throughout your body, like in between organs, and then there's the omentum fat that lies behind your intestines. So you can likely do something about your love handles, but your waist size may be determined by the inner fat. Yes, the FasciaBlaster® will enhance your weight loss efforts, but it can't do it all alone. At the end of the day, you have to manage your eating and exercise. AshleyBlackGuru.com/TheCelluliteMyth

3. <u>It cannot correct your posture</u>: As you learned in the last chapter, if you can't even stand in proper alignment, how in the world are you going to move in proper alignment? Posture is so important to overall health. While the FasciaBlaster® will open tight, restricted tissue and allow you to actually get into proper alignment, you have to neurologically reinforce the holding pattern or you are going to continue to lean, slouch, or twist, and continue to jack up your fascia.

4. <u>It has a limited effect on varicose and spider veins</u>: The reports are all over the board on the effects on veins. I'm not saying we can't get rid of the bulge or appearance, because people are reporting that they have. We do know that tight fascia causes disruption of venous flow, and when you support venous flow, it affects the appearance of veins. We've had some cases where they

disappeared, some where they improved, and some where there was no change. Set your expectations accordingly!

5. It has a limited effect on scar tissue: As I said before, scar tissue and fascial adhesions have the same composition. When we break up the scar tissue, it always tries to return, so you have to continue to regularly work on it. For example, my hip has so much dysfunction and scar tissue from my many surgeries. No matter how much I blast it, it is constantly re-forming to protect my dysfunctional, artificial hip joint. On the flip side, some people who were diagnosed with carpel tunnel get rid of it and it never comes back. Bottom line, scar tissue is broken down and reoccurs on a case by case basis.

6. It has limited results without heating: I'm about to tell you a whole lot more about this, but you need to know that your results will be limited if you aren't heating the tissue before blasting. There is a difference between what the physical structure will do in a cold state vs. a warm state.

7. It has a limited effect on skin tightening: Just to be clear, yes, you can tighten the skin to its best possible condition, be it on your face, neck, stomach, underarms, or so forth. This can be a long process, up to a year or more if someone has lost a lot of weight and has extra skin, or has a baby belly, and so on. You have to be realistic. Age also has something to do with it. If you are skinny and have loose skin, the FasciaBlaster® will tighten it to a point but you need to build some muscle to fill in the loose skin. The point is, keep working on it!

8. It has a limited effect on stretch marks: Here's another one where anecdotal responses have been all over the board. We don't know what the limits are with stretch marks. Once again, proper blood supply to the area has a huge impact. But there are clearly other variables. Just work on it and see what happens! Some people say they go away, some say they improve, and some don't see a change. My personal experience is that my "Mommy Marks" bruise when I blast and their appearance keeps reducing.

Remember, fascia restoration happens layer by layer. We're about to get into the protocols, but I don't want you to be chomping at the bit so much that you miss this important fact. Yes, most people see or feel or experience some sort of impact immediately, but your desired result may take a while. My friend Janette, who wrote the foreword of this book, is one example. She blasted and blasted and hit a wall with her results. Things just weren't improving. Then she used the tip of the Mini 2™ FasciaBlaster® that helps "dig out" divots or rake through larger chunks, and she worked with it to restore some more tissue and even developed a huge rash. Who knows what the rash was from! It was probably some sort of toxin that was trapped in the fascia that was being released and moved out of her body. She got through that and kept blasting. Then she won the chance to have a Fasciology session with Bart. She had another breakthrough in her body and is still improving little by little. The point is, we are complex and our bodies are complex, so we have to stay in the process until we see the breakthrough. It's a lifestyle change to get you to where you want to be. If one technique isn't working, try another. Don't give up. You are worth the time and effort! Please see additional resources at AshleyBlackGuru.com/TheCelluliteMyth.

8

Yaaaassss!! The Protocols

"Be happy for this moment. This moment is your life."

—Omar Khayyam

#Yaaaassssss!! The chapter on the protocols! I know you're thinking; FINALLY, right? But, you will soon understand why I needed to lay out all that groundwork for you. You can't give your kids the keys to the Ferrari until they go to driver's ed. You made it through fascia driver's ed and now here's your shiny prize!

First, let's talk about what you need. As discussed in chapter 5, there are some manual ways to address the fascia and reduce the appearance of cellulite; however, the FasciaBlaster® is hands down the most effective tool for the job. Every detail—the length, weight, size, and shape—were carefully and methodically thought out. I would be doing you a huge disservice to recommend anything else. It would be like telling you to walk from Jersey to L.A. when you can take a jet plane instead. Yes, I'm selling the "jet plane," but I invented it! Do you think Thomas Edison *didn't* tell people about the light bulb? It works! Okay, you get my point. MAJOR WARNING: DO NOT go trying to get creative using things like a garden rake or a meat cleaver. Other makeshift "tools" can damage the tissue and cause more internal scarring. Trust me, on this one. The FasciaBlaster® was designed to address fascia; a garden rake was designed for leaves. Unless you've got an elm tree growing out of your leg, leave the rake alone. Got it?

So you need a FasciaBlaster®, some oil, and a heat source.

STEP 1: PREP THE FASCIA WITH HEAT

The very first thing you want to do is heat the fascia, both internally and externally. Simply put, to internally heat, the blood needs to be pumping, which means you need to get moving! External heat means the temperature outside your body is higher than your body temperature, and we have a lot of ideas about how to achieve both.

Ideally, you would do at least 20 minutes of cardio, then spend 10 to 20 minutes in the sauna or steam room. And stop right there if you are starting to get into crazy thinking about how much time this is going to take. There are modifications you can make to fit into your schedule; in fact, the FasciaBlaster® works without heat, but it's a much slower process and can be more painful. Look at the women who have gone before you and they will say HEAT AND TREAT™! This protocol is the best-case scenario and truth be told, you should find an hour a day to spend on yourself every day. It's not always easy, we're all busy, we all have commitments, we all have obligations. But, we also all only have one body and one life. Love it and take care of it! (**#EndRant**) Back to heating.

When you heat the fascia, it actually changes states and becomes more pliable. Have you ever molded clay or even Play-Doh®? It's more malleable after it's warmed in your hands, and the same is true for your body tissue. But it's not just about the tissue becoming more malleable; it's about putting the tissue in a state for optimal healing and restoration. When you internally and externally heat any soft tissue structure, it causes physiological changes both locally and systemically. For example, if you heat your quad, changes happen both in the quad and throughout your body.

Ella Kim
January 11

I have fibromyalgia as well as Lupus and a host of other medical issues. I had been very sick for a while before I got any kind of a diagnosis. I spent roughly two years pretty much bedridden and felt isolated, lonely, and depressed because I couldn't be the mother I wanted to be.

In this condition, I could no longer help my husband run our business. My life was passing me by while I was stuck in bed or the couch due to the pain and exhaustion. During this time, I also developed cellulite. Terrible cellulite and it was something I had never dealt with before. The only reason I ever left the house was for a doctor's appointment. One day while I was online, I saw the FasciaBlaster® on my Facebook® newsfeed and was curious. I watched Ashley's YouTube® videos and then joined the group to learn more.

With the help of the FasciaBlaster®, I was able to come off all pain meds and anti-inflammatories. I finally found a tool that decreased my pain and improved my circulation. My cellulite issues are getting better each day and I am on my way to the fabulous legs I had before illness struck me.

I am so happy I found Ashley Black and the FasciaBlaster®! I'm back on my feet again. I have energy and no longer feel like my life is out of my control. The blaster has helped me take back the reigns and regain my confidence. It gave me the hope I desperately needed and helped me to feel like me again. Thank you Ashley Black!

Here is what's happening at the site when you internally and externally heat:

- Vasodilation, or opening of the blood vessels (This is good!)
- Increased nerve conduction velocity, meaning nerves fire faster (This is good!)
- Increased metabolic rate (This is *very* good!)
- Increased capillary permeability, which is the transfer of oxygen between veins and arteries (This is good!)
- Increased leukocyte delivery, meaning more white blood cells come to the site and help with cellular repair and waste removal (This is good!)
- Increased collagen extensibility, or more range of motion allowed by the collagen fibers (This is good!)

- Increased venous and lymphatic drainage, which is improved blood flow, detoxification and immune system response (This is good!)
- Decreased muscle tension and spasm (This is good!)
- Loosened fascial "grip" on muscles, meaning they will be more penetrable by the FasciaBlaster® (Basically, you can get in there!)

As you can see, lots of good things happen when you heat the tissue. Additionally, when you heat your entire body, such as through a warm whirlpool, steam room, sauna, or the Texas sun, it increases your metabolic, pulse and respiratory rates, and it decreases blood pressure.

Internal Heating

Internal heat is exactly that—the body's own way of heating itself. This is done by cardiovascular exercise, which means you are working your heart. Now when I talk about cardio, first of all, let's make sure you aren't doing something that is going to jack up your fascia.

A lot of people jump into intensive training programs trying to get a quick result, but they get a quick injury instead, or fascial recoil. This is because the body wasn't ready for what you asked it to do. So, let's slow the boat down and use some common sense. (I should just say "sense" because unfortunately throughout the industry, it just isn't that common.) When you do any kind of impact exercise, such as walking, running, or jumping of any kind, a lot is happening in the brain and body. Remember our brain to body communication system we talked about in chapter 3? Well, there is a lot of communication happening every time you pick your foot up off the ground. The brain is rapidly firing trying to figure out where and how to put the foot back down. At the same time, the fascia has a proprioceptive mechanism, which means it's also sensing the movement and trying to figure out if it's healthy movement or damaging movement. The fascia wants to do its job properly and it's trying to decide if it should clamp down to protect the body, or if it should stay open and allow that movement. Crazy, isn't it? And you thought that when you exercise, you are clearing your head!

If you had a perfectly aligned body (knees, hips, feet, shoulders, posture, etc.) and your movement patterns were 100 percent perfect, then *any* exercise would be fine. However, if—like 99 percent of us—you have any deficiency in your body, that's when you have to really look at what you are doing from an exercise standpoint. It's so important because the fascia senses the deficiency and tells the brain, then the brain tells the fascia to protect, and unfortunately protection comes in the form of a lockdown. So activities that are risky to the fascia are walking, running, kickboxing, rebounding, dancing, and so on, any movement where your feet come off the ground. This is exactly why "fit" people have some of the worst fascia and cellulite. Remember the section on the brain to body connection in chapter 3? This is why you have to lay the ground work before doing anything like CrossFit®, P90x®, Pilates, Yoga, boot camp, squats and lunges. I know, I just rattled off everything you've been told to do for the last decade. (**#AnotherBook**) If you have structural deficiencies at all, these are your high-risk movements. Don't shoot the messenger. I warned you in the beginning that I was going to dispel a lot of myths and give you the TRUTH!

Before doing any type of complex movement, you have to prepare your body structurally and mechanically. (See page 76.) People get injured because they are simply not ready for the level of exercise they are choosing. I also believe this is one of the reasons that it's so difficult for people to stay on an exercise program even if they are fit. Your own body is subtly and subconsciously begging you to stop because it's not ready for what you are trying to force it to do! Even some pro athletes are not ready to run efficiently, so get it out of your head that a person's fitness level has anything to do with being "fascia-ready" or being "structurally and mechanically" ready.

Now, I am certainly not trying to keep you from exercising, but I am trying to make you aware that if you don't have a proper structural and mechanical foundation, if your body is out of alignment at all, your fascia is going to keep clamping down on you. This is why most people who are Beyond Bound™ are hardcore exercisers.

What type of cardio should you start with? Your low-risk movements are movements where your feet stay in the same place, like using a stationary bike, an elliptical machine, The Gazelle®, or even single-joint muscle training. Once you get into the right posture, the fascia doesn't need to make any more decisions or figure out how to protect the body, because there are no surprises in your movement patterns. The brain will allow the fascia to relax.

If you have any questions at all about how you move or how to neurologically align your body and prepare for more complex movements, all of this is found in my exercise videos at AshleyBlackGuru.com/The-CelluliteMyth. Once you learn how to properly move, you can transition into anything else you want to do and your fascia will quit attacking you. However, if you insist on continuing the high-risk movements without a proper neurological foundation, you're going to have to use the FasciaBlaster® just to try to maintain normalcy and you'll have to use it extra to get ahead and get rid of cellulite. Don't take two steps forward with fascia restoration and two steps back with bad posture and high-risk exercise. Save the exercise for once you've restored your structural and mechanical integrity. This is a huge part of taking your body to a level it has never been before. Just ask our fitness stars!

External Heating

External heating practices for health date all the way back to ancient times and are a part of cultures all around the world. In fact, not long ago I attended a communal bathhouse in Japan, called an "Onsen," and it was absolutely one of the most beautiful experiences. Women were nurturing and caring for one another, older women mentoring the younger women. Each woman had such confidence. This is the way it should be, ladies! No competing, no comparing. Just love and acceptance among us.

I'm sure this is the way it was in Ancient Rome too. Ancient Roman bathhouses were the center of society, and they weren't just for cleanliness; they were for health. These communal bathhouses varied from simple to elaborate architecture and contained a series of rooms with different baths where the water got progressively hotter. The bather would start in a cold bath, and after progressing through each warmer bath and breaking a sweat, the bather would again dip in cold water. Next, the bather would enjoy a massage with oils and a final scraping with metal implements—a rudimentary form of fascia treatment! While this isn't the protocol I would recommend with the knowledge that we have today, it's amazing that what they were doing is so incredibly close to my protocols. It's even more amazing that somewhere along the way we lost this discipline—but it's time to bring it back!

While I'm a huge fan of the cultural and health benefits of heat, my favorite way to prep my fascia through external heat is in my sauna, primarily because I can simply blast while in there. Here is a list of my other favorite ways:

- Steam room
- Hot bath
- Sauna suit
- Zip-up sauna
- Heating pad

Again, you want to heat for at least 10 to 20 minutes. A good guide is that you want to stay with it long enough to break a sweat. And if you are really an overachiever (like me) you want to actually do your blasting in the heat. Of course, you really can't blast while wearing a sauna suit so make adjustments accordingly.

Monica Dickens
June 3

I'm hands down a believer!! Forget cellulite reduction. If that's a perk then I'll take it. After years of being a high school/college athlete, and several years of Crossfit, this tool is a godsend! I blasted for the first time this morning and hit spots on my legs that I NEVER suspected to be bound or tender! I've been going to a chiropractor and getting biweekly massages now for almost 2 years and have never had the release that I got from 1 session of blasting!

STEP 2: OIL UP!

Once your tissue is heated, it's time to choose the area of your body that you want to work. This is totally up to you, sista! Remember all that data you collected when you were analyzing your fascia? Where is your cellulite? Where are your fat pockets? Where are the spots that you can't pinch? Where are the places that showed distortion when you poked? Where are the places that were tight when you tried to stand in correct posture? Where do you have pain? Those are the areas to really focus on at first. Or, simply start with the area you want to improve the most visually. It's your body, so it's your call.

For sure, you NEVER want to blast on dry skin or through clothing, so you need to bare your skin and get some oil. (Sounds fun, doesn't it?) You can use coconut oil, massage oil, or even olive oil if that is all you have. But for the best blasting experience, I developed Blaster Oil™, which was formulated at one of the top labs in this country. First of all, it smells dope! I always want my products to reflect who I am as a person, and since I love the beach, my Blaster Oil™ has the most subtle and amazing beach aroma! But it also has some other features of "Ashley Black-ness." Blaster Oil™:

1. Contains proprietary ingredients that activate the fat-burning mechanisms of adipose tissue. This means cellulite is reducing before you even start blasting!

2. Contains ingredients that stimulate circulation which accelerates healing and fascia restoration.

3. Is thicker than most oils so it's slow-absorbing, which means you don't have to keep reapplying as you blast.

4. Comes in a convenient spray pump so your hands don't get greasy, which will impact your grip on the FasciaBlaster®.

As you are blasting, you are opening up the tissue and my Blaster Oil™ will gobble up the fat cells as you go. Of course, my oil would be my top recommendation since it's specially formulated for this process; however, as I mentioned before, you can use any oil to get started—even 99-cent baby oil! (Just be aware there is a high correlation between baby oil and rashes.)

Applying the oil to your skin shouldn't need much explanation, but let me just say, if you apply the oil with your hands, they will be slick when you try to hold your FasciaBlaster®, which can be really frustrating and impact the effectiveness of your session. This is why I offer my Blaster Oil™ in a spray bottle so you can keep your hands dry. Whatever oil you choose to use, apply the oil directly from the bottle to your skin and then use the FasciaBlaster® to spread it around on the area. Make sure you have enough oil on your skin for the FasciaBlaster® to glide along the top of the skin and not tear or irritate it. More is better than less. Use oil generously!

STEP 3: GET YOUR BLAST ON

Now that your fascia is properly heated and all slicked up, it's time to get to work! Your first goal when addressing your fascia and your cellulite is to break up the Hail Damage that is a result of the condition of the first layer of Ace® Bandage/Structural fascia. If you do not have Hail Damage, you would still begin blasting lightly on the chosen area to stimulate blood flow and see how your body responds to the process. In the next chapter I'll get into some questions you may already have but for now, just familiarize yourself with the process and go easy on yourself until you see how your body responds. Some people really start to detox when they start blasting so just be aware. Use fast strokes with very little pressure. Never use the FasciaBlaster® slowly or it won't work. To blast, you are going to lightly "scrub" the fascia, keeping the following in mind:

a. **Pressure**: Lay the claws of the FasciaBlaster® on the skin and push down until you are just on top of the muscle, or where you feel the first resistance pushing back at you, or about ¼ inch deep. To address Hail Damage, pressure should be fairly light, or about the weight of the FasciaBlaster® itself. Don't press into the skin too deeply.

b. **Stroke**: Long strokes are most beneficial. You can blast with or across the muscle fibers. Go up and down, side to side, but never in circles and never weaving. Do not try to massage with the FasciaBlaster®; you will not get results. The FasciaBlaster® is not a massager, it's a fascia tool.

c. **Speed**: This is very important. You want to go as fast as possible, almost like a cheese grater, approximately two reps per second. The reason you want to go fast is because you are heating the tissue with friction and naturally loosening the fascia by stimulating blood flow. Also, the fast motion breaks apart the little adhesions. If you do it slowly, the claws of the FasciaBlaster® will just bump and roll over the adhesion. You want to hit it hard, fast, and continually to see results! It's sort of like getting a tangled knot out of your hair. If you try to comb through all of it, it's going to hurt and you're going to rip out your hair. Instead, you start on the outside and work the tangle a little at a time. Your fascia is tangled up in a similar way. Do a little at a time.

d. **Duration**: Blast for about 3 to 5 minutes per body part or until you feel a slight tingling or the skin is warm, or until you see some redness. Any single area may be worked on for up to 10 minutes. Now, it may take a month or more for you to work up to this amount of time. Blasting is a workout and some people start at 1 to 2 minutes. That's okay. If you only want to focus on your trouble areas (where you have the most "dimples" or crappy fascia) you can work on those spots every day for up to 10 minutes per area. (See the following pain section for exceptions.)

e. **Pain:** Tight fascia will be sore, but as you blast it and it opens, the pain goes away. Healthy tissue is not painful to the touch. Wherever there is pain indicates a need to open the fascia. On a pain scale of 1 to 10, never go above a 7 when you are blasting, and a 5 is preferable. Too much can cause swelling and actually take longer to restore. It can also cause a fascia recoil, which will set you back a bit. Just take it easy and work up to longer and higher tolerance. If you are raging in pain after one minute, or hitting an 8 to 9 on the pain scale right away, then only do a minute and try it again the next day. Ultimately, you know the fascia is open in an area because it simply doesn't hurt anymore and it looks smooth and supple. You can also measure your progress by going through the 5 Ps of analyzing your fascia in chapter 6. For more information on pain and bruising, see page 168.

f. **Frequency:** You can blast every day; however, recovery time is important. Don't blast over areas that are swollen or sore. You can blast over bruises if they are not sore. Initially, because the fascia is being released at the very surface, bruising will be more prominent than when you move on to treat deeper layers of fascia. The deeper you go, the less bruising you see. Ideally, you should blast for about 30 days and then take a week off and you should see some amazing results! (Make sure you have those "before" photos.) Then repeat the cycle up to 90 days. You might ask, "Can I do legs one day and upper body the next?" The answer is yes. You have to fit whatever you can into your schedule and lifestyle. My suggestion is to do every body part for 3 to 5 minutes every day. It will take a little longer to get your results if you skip days but you'll still get there! Blasting is an art form and you have to find your groove. Everyone's pain tolerance is different so take these as general guidelines. And to be totally honest, with thousands of women reporting their blasting routines, everyone does it just a little differently. My girlfriend advice—JUST BLAST!

29 BLASTING ZONES

For best results, make sure you are blasting your entire body because the fascia system works as a whole. This might mean you do half one day and half the next, targeting your trouble spots every day. Here are the 29 "zones" to blast on your body:

1. THE FRONT OF YOUR RIGHT THIGH

Blast toward the hip

Blast toward the knee

Blast side to side

2. THE FRONT OF YOUR LEFT THIGH

Blast toward the hip Blast toward the knee Blast side to side

3. THE BACK & SIDE OF THE RIGHT THIGH

Blast back side of thigh   Blast back side of thigh   Blast side to side
    toward the hip   toward the knee

4. THE BACK & SIDE OF THE LEFT THIGH

Blast back side of thigh   Blast back side of thigh   Blast side to side
    toward the hip   toward the knee

5. RIGHT ILIOTIBIAL (IT) BAND

Blast side of leg toward the hip

Blast side of leg toward the knee

6. LEFT ILIOTIBIAL (IT) BAND

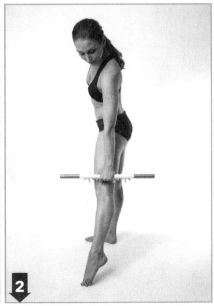

Blast side of leg toward the hip

Blast side of leg toward the knee

7. INNER RIGHT THIGH

Blast toward the inner right thigh   Blast toward the knee   Blast across your right thigh side to side

8. INNER LEFT THIGH

Blast toward the inner left thigh   Blast toward the knee   Blast across your left thigh side to side

9. RIGHT LOWER LEG

Blast toward the knee   Blast toward the foot   Blast toward the knee   Blast toward the heel

10. LEFT LOWER LEG

Blast toward the knee   Blast toward the foot   Blast toward the knee   Blast toward the heel

11. RIGHT SADDLE BAG

Blast toward the hip   Blast toward the knee   Blast across the side of the thigh below your hip, side to side

12. LEFT SADDLE BAG

   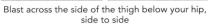

Blast toward the hip   Blast toward the knee   Blast across the side of the thigh below your hip, side to side

13. RIGHT GLUTE/HAMMY TIE-IN

Blast right below your right buttocks, side to side

14. LEFT GLUTE/HAMMY TIE-IN

Blast right below your left buttocks, side to side

15. RIGHT GLUTE

Blast up your buttocks   Blast down your buttocks   Blast across your buttocks, side to side

16. LEFT GLUTE

Blast up your buttocks   Blast down your buttocks   Blast across your buttocks, side to side

17. STOMACH

Blast across the stomach, toward hips

Blast across the stomach, toward chest

Blast across the stomach left to right, diagonally toward hips

Blast across the stomach right to left, diagonally toward hips

Blast the pubis diagonally, right to left

Blast the pubis diagonally, left to right

18. RIGHT LOVE HANDLE

Blast love handle, toward chest

Blast love handle, toward hip

Blast love handle, side to side

19. LEFT LOVE HANDLE

Blast love handle, toward chest

Blast love handle, toward hip

Blast love handle, side to side

20. RIGHT ARM

| Blast side of the arm toward the shoulder | Blast side of the arm toward the elbow | Blast triceps toward the elbow | Blast triceps toward the shoulder |

Blast front of the arm toward the shoulder   Blast front of the arm toward the elbow

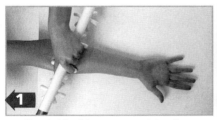

Blast inside of forearm toward the elbow   Blast inside of forearm toward the hand

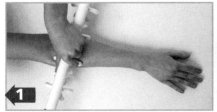

Blast top of forearm toward the elbow   Blast top of forearm toward the hand

21. LEFT ARM

Blast side of the arm toward the shoulder

Blast side of the arm toward the elbow

Blast triceps toward the elbow

Blast triceps toward the shoulder

Blast front of the arm toward the shoulder

Blast front of the arm toward the elbow

Blast inside of forearm toward the elbow

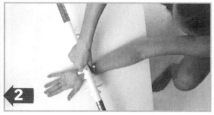

Blast inside of forearm toward the hand

Blast top of forearm toward the elbow

Blast top of forearm toward the hand

22. CHEST

Blast across the chest, side to side

Blast diagonally across the chest, left to right

Blast diagonally across the chest, right to left

23. RIGHT NECK

Blast right side of neck toward the ear

Blast right side of neck toward the collar bone

Blast across your right neck, side to side

24. LEFT NECK

Blast left side of neck toward the ear

Blast left side of neck toward the collar bone

Blast across your left neck, side to side

25. BACK OF NECK

Blast the back of the neck toward the top of head   Blast the back of the neck toward the shoulders

26. HEAD

Blast from the forehead to the back of the head (best when done in the shower with conditioner)

27. HANDS

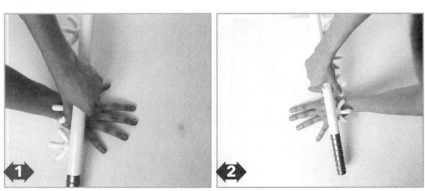

Blast the top of the right hand   Blast the top of the left hand

28. FEET

Blast the top of the right foot

Blast the top of the left foot

Blast the bottom of the right foot toward the heel

Blast the bottom of the right foot toward the toes

Blast bottom of the left foot toward the toes

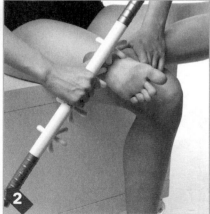

Blast bottom of the left foot toward the heel

29. SPINE

Blast the midback
toward the shoulders

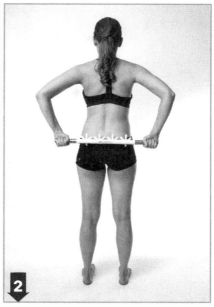

Blast the midback
toward the buttocks

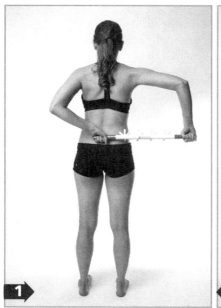

Blast across the spine, side to side

What Not To Blast

1. So, this may come as a surprise, but The FasciaBlaster® is for EXTERNAL USE ONLY. And that's all I'm going to say about that. However, you can blast externally around down "there." See the next chapter for more information on the pelvic floor.

2. Do not blast vigorously over your lymph nodes. It is unlikely to cause a problem but as a precaution, just use some common sense. See page 177 for description of where the lymph nodes are located.

3. Do not blast breast tissue. You can blast around the tissue, including lightly over the skin of implants or scar tissue but do not blast the actual tissue. Breast tissue is chemically different from fat and we just don't know yet the effects. You can use it for "pit tits," which is the little bulge that's between your arm and your breast. (One of my personal favorite places, being a big boobied gal.)

4. ABSOLUTELY NO BLASTING if there are blood clots or if you are on an anti-coagulant blood thinner. See the next chapter on contraindications and risks.

STEP 4: AFTERCARE

In an ideal world, you would end the protocol with a cool shower, cool bath, or an icepack. If you are really having intense soreness or bruising, throw three bags of ice from the convenience store into the tub and fill it with enough water to immerse yourself. Not only will this help with inflammation, but it also is reasonably common knowledge that ice baths trigger the burning of another type of fat, brown fat. Ladies, athletes do this all the time and we should too!

Then after you cool, apply a restorative cream. My personal favorite is my After Blaster™ Cream, which has three special ingredients formulated for this process:

- Arnica to promote healing and remodeling the fascia
- Skin tightener

However, if you already have some sort of topical support that you love, like arnica gel, or essential oils, use it. It's just a nice way to finish off the protocol and make sure you maximize your efforts.

Finally, we've talked about the importance of hydration on page 80 so make sure you get some Cellectrolytes™ to keep yourself hydrated and balanced. Ideally, you would take them throughout your blasting session.

GOING TO THE NEXT LEVEL

In the last chapter, I talked about what kind of results you might be able to expect from the FasciaBlaster®, but I want to briefly touch *how* to expect it, or the way it all might come together for you. Just know that the process is going to look different for every single person because our bodies and life histories are so different. What happened for one person may not for the next the same way. That's why I created my FasciaBlaster® Facebook® group that had over 100,000 women at the time this book was first printed. Seeing pictures and reading testimonies will help you understand your own journey so you can stay on track. Staying connected to the group is a *huge* piece of the success puzzle because as humans, we just do better when we have support on any task. If you're not a member of my private, women-only FasciaBlaster® Facebook® group, go join right now! You are about to begin a life-changing processes, so don't go it alone (if you haven't already started). I cannot teach you an entire science in one book, but I post and teach in the group daily. That's why my brand's tagline is "Knowledge. Empowerment. Inspiration." Go learn, get yourself figured out, and then pay it forward. **#TheAshleyWay**

Ultimately, only you can decide when you are ready to get into the deeper layers of fascia by blasting harder or using my other complementary tools and products, such as the nugget, which helps "dig out" divots or rake through larger chunks. Go to AshleyBlackGuru.com/The-CelluliteMyth to learn about advanced techniques. Just start learning about your body! It's amazing that we can live in a body for 30, 40, or 50 years and not really know it!! Get ready to **#KnowThyself** because this process will help you see yourself in a whole new way!

MEASURES OF SUCCESS

You will have many changes in your body over time, and while getting rid of your cellulite may be your most important goal right now, don't let the appearance of your skin be your only measure of success. You might simply feel better or get rid of some random pain before your cellulite disappears. Just know that it's all part of restoring the fascia to a more healthy state, so really pay attention to what's going on. Is your sleep better? Are you detoxing? Are you feeling more focused? Are you feeling muscles you've never felt before? It's all related to your fascia. That's why the 5 Ps are so important because when you re-check, you might see a drastic increase in your range of motion, or some other *awesome* benefit before you see your cellulite disappear.

I explained some of this in chapter 3, "Healthy Fascia vs. Unhealthy Fascia" but biomechanically speaking, bad fascia is like putting on clothes that are too tight or improperly made. If you put a shirt on with one arm hole sewn shut, you can't possibly wear it and function optimally. What can you do with one arm pinned to your body? That's perhaps an extreme example but tight surface fascia can pull you out of alignment and restrict you anywhere. It's the reason a person can have a chiropractic adjustment and then the next day be right back where they started. The tight fascia pulls the body back out of alignment. You have to address both tissue and moving in alignment at the same time to get permanent results—which takes us back to **#StructuralIntegrity** and **#BiomechanicalIntegrity**. We are only able to move within the confines of what our structure allows. By opening the fascia, the structure can adjust itself and movement will naturally improve.

For example, one of my clients (who then became my co-author, but we won't mention any names) was so structurally restrained by tight fascia that she couldn't even do a proper lunge. Her body just wouldn't let her get into the right position but she was oblivious to the structural restriction and mechanical dysfunction. She was "faking it" and thinking she was doing it right, like many of you are doing. After one hour with me, she was able to not only do a proper lunge, but she did it without pain for the very first time in more than 30 years. Not only had her movement been restricted, her muscle access was restricted as well. Opening the tissue impacted her on multiple levels. (Yes, she cried. Happens all the time. More on emotions in the next chapter.) This is a common thread with my clients and followers—the impossible becoming possible, the hopeless finding hope and solutions, and lots of crying! **#HappyCry!** Best Ah-ha moment reactions evah! I live for this stuff!

Anyway, I say all of this because I want you to be prepared for changes that are going to take place in your body. If you've had atrophied muscles for a long time and you open those muscle fibers, you want to make sure to activate and strengthen the muscle so that the fascia doesn't tighten back down to stabilize it. There are lots of ways to do this, but the no-brainer **#EasyButton** are the Ashley Black exercises, which are exercises designed to help you create a foundation for proper movement. Visit AshleyBlackGuru.com/TheCelluliteMyth for more information.

Keep in mind that as you go through the layers of fascia, each layer itself has to go through a process of restoration. If you are Beyond Bound™, you are going to go through the longest process of change. Your skin might start to look physically more dimpled while it is being restored, but keep going. You might find that you have a lot of chunks that were so tight that they appeared to be smooth at first, but now the chunks are breaking up and they feel bumpy. Keep going to get the smaller chunks smoothed out. And DON'T PANIC. There's no chance that you'll ever be stuck in a phase or a type. The FasciaBlaster® ends all of it over time with the proper understanding and the proper techniques. I also have a radio program called "Body Hacks" on my Billionaire Healthcare® radio show that will help you understand more. Visit AshleyBlackGuru.com/The CelluliteMyth to learn more.

You probably have a ton of questions right now that will hopefully get answered in the next chapter. Take a deep breath and trust your body to guide you along the way. Listen to your body and appreciate how truly amazing it is!

Sherry D Martinez
April 22

Take it from me, Sherry D . . .

"Fascial distortion free" is what you want to be!

Who knew, cellulite is a symptom and not a curse?

Who knew that freeing your fascia could "blast" your life onto a journey that you never imagined? That you could free your mind and body and soar to new heights that you used to only dream about?

What is Fascial distortion you ask?

Ashley Black's dedication, education and passion for the human body and biomechanics has led her to the deepest, most profound, history-making discovery—the ability to manage and manipulate cellulite, for a smoother appearance and lyse fat cells away!

I too, like Ashley Black, have had a life-threatening illness, that kept me bedridden for months, not knowing if I would ever be able to dance or exercise again. I, like Ashley, turned to self-treatment on my own body, due to structural and internal issues of my own. I was blessed to have come across the FasciaBlaster® online one day and as soon as my eyes met those tiny little claws, I was determined to get my hands on one. I knew instantly that this was going to be the tool, that I needed to reach deep within my "soft tissue" (we now know to be fascia), and break up the huge adhesions that were growing within my body at a rapid rate, causing my muscles to atrophy.

I bought the FasciaBlaster® for pain management to break up the fascial adhesions in my body that were causing muscular debilitation. That worked within the first few treatments! As I watched the FasciaBlaster® Facebook® page day in and day out, I was seeing women post pictures of their bodies changing in appearance and shape. So, I thought why not give it a try. When I learned and discovered how to fat lyse and pop fat cells I was hooked. As I sculpted my body, I discovered something much more profoundly deeper!

I felt as if freeing my fascia has also freed my mind. I am able to see old issues that I was carrying around as blessings rather than baggage. As I scrub away my saddlebags, I wash away all of the mental "yuck" that had been clouding up my life.

I can offer as a witness, once a week, scalp to toe FasciaBlaster® treatments in a warm Epson salt bath, will detoxify old trapped fungal issues that cause fatigue and skin rashes. My energy has increased, my skin rashes are leaving my body, my appetite has increased, due to the lean muscle mass that I now have:).

I have shrunk in size, I now have a body that resembles that of a swimmer, long and lean 50-year-old FasciaBlasting machine! The more freely my body flows, the more free I feel—free to achieve anything I set my mind to. Yes, 50 looks good and feels great. I am now fascia fabulous!!

9

The Forthcoming Fa-scia-it Storm

*"Before anything else,
preparation is the key
to success."*

−Alexander Graham Bell

Well, congratulations! If you made it to this point in the book, you've probably taken the "Black" pill and well, there's just no going back! You either are already experiencing radical changes in your body or you are about to. Either way, there's no doubt you have a *ton* of questions, and that's a good thing! You are thinking for yourself and taking control of your health. Ask away! There are no wrong questions and hopefully this chapter will answer most of them. Of course, if you don't find the answer you need, message my team. Also, check out the resources at AshleyBlackGuru.com/TheCelluliteMyth. That's what they are there for! But please, do your due diligence and read this chapter first and thumb through the pages you've already read in this book. We really tried to be as comprehensive as possible in this chapter. You may even remember when I posted on the group page and asked for topics you'd like to see covered. It's all a bit random, but it's all related to blasting. So, consider this chapter like a factoid stew. We threw all of this information together because, well, we just really didn't know what else to do with it—but we promise it will be delicious! (With a side of cornbread.) You might want to read this chapter with a highlighter and mark what pertains to your situation.

You may be wondering a little about the title of this chapter. It's a bit tongue-in-cheek but be forewarned, there's a lot that goes on in your body when you start blasting. Questions will come up as you go. Changes may happen that you weren't expecting. I know. I've been there and I've coached thousands upon thousands of women through this process. You aren't some weird exception to the rule who can't get results. Everyone who does this properly and consistently sees results . . . everyone! It may take you longer than one of your friends but stay the course. It's worth it, I promise! Slow and steady wins the race, crazy ladies!

THE NEWS ON THE BRUISE

Anyone who starts blasting knows that even though *you* feel amazing, the people around you may be giving you strange looks. That's because restoring fascia can mean bruises. Not always, and not everyone gets them, and they may not show up right away. You may have an awesome blasting session and feel super great, and then a few hours later, or the next day, you have all these tiny black and blue marks. Or, you may have a giant purplish strawberry across your whole leg. Don't freak out. As bizarre as it sounds, it's completely normal and actually a good thing.

The bruise often gets a bad rap. I feel like we need to invent a phrase for "good bruise." Yes, it's mostly associated with pain and injury, but bruising means healing. It's sort of like "fever," which also gets a bad rap. Fever is your body's defense mechanism to an invader. Fever is good! Of course, too high of a fever is not good, but at its core, fever is simply an immune response. Same idea with bruising—it's a healing response. It is nothing more than blood rushing to the scene to heal and restore the tissue that has just been opened. Fascia restoration bruises are healthy, restorative, and cleansing.

Here's why you bruise when you are blasting:

- When you are smoothing out a fascial adhesion, it's like pulling two pieces of duct tape apart. There's going to be some inherent bruising that comes with that type of restoration. It's a sign that you've successfully separated the sticky layers of tissue. Blood rushes to the area so that healing can begin in that area.

- When someone has restricted fascia, it's like having a kink in a garden hose. When you open the fascia, all of the sudden you have blood flowing to that area again. These types of bruises are not painful but they can be immediately visible. They can also be accompanied by a tingling or itching feeling as blood rushes through the tissue.

Emi MacLeod-Loop
December 27

I ordered the FasciaBlaster® on a very skeptical whim, based on some Internet searches I had done for my hip that was rotated due to a difficult childbirth. I started seeing all these pictures of women who had crazy bruises on their limbs and torsos but they had these amazing results of fat loss, cellulite loss, smoothing of skin as well as therapeutic help for all kinds of pain issues. It seemed like a longshot for me but I went ahead and ordered, even though I told myself that this was yet another crazy . . . shopping scheme . . . scam . . . "thing" that I was trying to do to fix myself.

Because of my rotated hip, I had horrible lower back pain and mobility issues. What I also had was a decade and a half of almost unexplainable weight gain no matter what I did. I ate healthy, I ate normal, I did the Atkins diet, I took shots of HCG in conjunction with the recommended very low calorie diet. Nothing provided any kind of long-term help. My body type actually helps me to build muscle rather quickly so I could never understand why I was always gaining weight without over eating.

I could also never understand why I had what seemed to me to be chronic fatigue issues, lack of motivation, and depression. I can honestly say that when I ordered the FasciaBlaster® I really thought this was just going to be another sinkhole for money that I really couldn't spare.

Within the first week of blasting, my lower back pain was almost completely gone. All of the sudden, it was like I had a new lease on life! I had more energy and my attitude towards life in general was improved. My sex life . . . for the first time in almost a decade my sex drive was back to normal for me, my husband and I discovered a whole new aspect of our relationship. Having intimate time was no longer something that I had to work to get in the mood for.

I am fairly new to blasting; I have almost hit the 60 day mark. In this time my cellulite has smoothed, I have lost major inches in my hips and belly overhang, upper groin area, and thighs, which are the areas that I have focused on in blasting. Ashley Black has literally changed my life. I really don't have the proper words to endorse the FasciaBlaster® or express what it truly means to me.

The best part is, the longer I continue to blast my body, the more I find that it's possible to reshape my body, my health, and my psyche into the image I have in my head of what I really look like, and how I should really feel. I am 38 years old with a 10-month-old baby and yet I have always been surprised when I see my appearance in a mirror because I have this image of myself the way I looked when I was in my late teens/early 20s. For the first time in a decade and a half, I actually feel like my internal image of myself can be matched with what I see in the mirror. I owe all of this to Ashley Black and her FasciaBlaster®. This is truly a miracle product and I can never say thank you enough.

"LUCY! YOU HAVE SOME SPLAININ' TO DO!"

Now, I know this is going to come as a major shock to you, but we've had many, many reports from users that their husbands, partners, and loved ones are not happy to see them covered in bruises. (I know.) How should you handle it? Well, there are a couple of approaches. First of all, remember, recovery time is important. While you may be a type "A" go-getter, you have to pay attention to your body. Take a break or don't blast as aggressively. And if you do bruise, be conscientious and make sure you wear proper covering when you are out. Nowadays, there are all kinds of covers and wraps that will conceal the arms without being too hot. You can also explain, "I'm restoring my fascia. The bruises are the blood rushing to the area and healing the tissue." And finally, you can give them this book! **#AshleyMadeMeDoIt**

You should also know that for as many reports as there are about husbands and loved ones freaking out about the bruising, there are reports of husbands and loved ones asking to be blasted! So, just remember that some people (men in particular) take a little longer to come around but 99 percent of the time, they do! People might roll their eyes or gasp in shock but remember, you aren't doing this for anyone else but yourself. No one else lives with your bad fascia day in and day out but *you*. No one else feels your pain or celebrates your gain but *you*. When others start to see the difference, when they see your gorgeous confidence and power coming through, they will be as supportive as anyone! So, BlasterSister™, stay the course, be kind, and blow their minds with the changes! **#BlastOn**

TALKING TO YOUR HEALTHCARE PRACTITIONER

Please, don't be afraid to talk to your healthcare practitioner. Be assertive! We've had lots of reports of interested medical professionals looking into fascia and many who use the FasciaBlaster® and are followers. Please, tell them you aren't being abused, because it's the law that they report abuse. Simply tell them you are restoring your fascia!

Misty and Cameron Fant
June 6

I got the pleasure of sharing with a plastic surgeon on the FasciaBlaster® today. He wasn't negative at ALL! He said you look great for having an 8 month old. What are you doing? I said let me introduce you to the FasciaBlaster®!! He said I need that for my belly so I go to pull up b4 & after pics of ladies bellies and he said are you a sales rep for them?! I said no but you REALLY get REAL results . . . buy one! FYI, His nurse assistant had it up on her IPad® scanning the website! #BlasterBeliever #thanksAshley

FINDING A BLASTING BUDDY/PRACTITIONER

So another thing I get asked All. The. Time. is, "Does anyone do this professionally?" I get it. I don't like to blast myself all the time. It's nice to have someone else do the work every now and then while you sit there gritting your teeth. You can get almost anyone to help you do this. BlasterSisters™ have united, in person, to help one another. Many licensed practitioners have also incorporated the FasciaBlaster® into their work. I want to make the FasciaBlaster® available to every man, woman, and child. We all need healthy fascia. MAJOR DISCLAIMER: Use these people or practitioners at your own risk!! These are not people I know personally, nor have I trained, nor have I (probably) ever even talked to!

Great news though: Stay tuned for our app to help connect you to BlasterSisters and practitioners. We are working on it...diligently.

"WORSE BEFORE BETTER" MYTH

If I were handing out awards for what issue causes the most drama on my Facebook® page, it's this one. But it's really not that big of a deal. In fact, I want to call this the getting "Better Before Better" phase. What's happening is that sometimes you have a false appearance of smoothness because the tissue is Beyond Bound™. Sort of like plastic wrap tightly secured over meat. When you release the plastic wrap, the meat spreads out and the shape changes. Or you could say it's like Spanx®. (I know, now I've gone to meddlin'!) Spanx® is "fake" tightness that makes you look smooth. The damage is hiding underneath. Beyond Bound™ fascia is like Spanx® under the skin. As you start to release it, the true story is told.

In your body, when you start breaking up those tightly bound chunks, you are going to get smaller chunks that might temporarily give you a more textured or "loose" look. But keep going! You are making progress! Think of it like renovating a house. During the demo phase it can get messy, but you have to demo in order to remodel. You are in a process. Keep going!

Here's what you can do to get through this stage a little faster:

1. Keep blasting LIGHT and FAST. NO digging.
2. Lightly weight train. You want to immediately fire the newly accessed muscle fibers and flush blood to the area, but don't overdo it. You are working brand new baby muscle fibers! Treat yourself like someone just coming out of surgery.
3. If you are working on your legs, do leg extensions and curls, abduction and adduction calf raises to isolate the legs.

BLASTING PETS

Yes! Pets have fascia too! In my early career I performed manual therapy on horses and learned they have similar issues as humans with muscles, joints, ligaments, and tendons. In fact, every animal has fascia and every animal has muscle and connective tissue. Blasting is very good for them, particularly around the joints. Also, fur serves as a natural lubricant so you can blast right over it, no oil needed unless their fur is minimal, in which case oil them up!

We intend to do further research in the future, but as long as they let you do it, go ahead and blast your pets. If you have any questions or concerns, please consult with your veterinarian about fascial release.

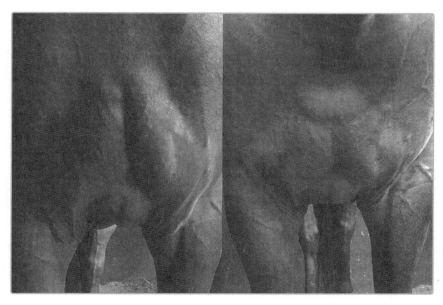

This image is of a horse's chest. You can see the tense muscles bound by fascia in the before picture on the left, and the smooth muscles in the picture after blasting. See the trainer's comments below.

 dragonbalanceequine
July 12

Following the suggestion of a fellow #bodyworker, I bought a #fasciablaster to use on the horses. I tested it out on #sirwilliam and the #dragon today, and saw some remarkable changes! These photos show the difference in Abs' chest. I've struggled for years to help her soften these muscles and while I've seen and felt positive results, I've never seen such a big change in one session. She holds most of her tension in this area, so I'm very excited to ride her after this release!

✛   💬   ✳✳✳✳✳

BLASTING KIDS

Fascia recoil can start as early as the womb. Our very first cells of life are fascia, and if there are any complications in the womb or in the birthing process, fascia can already be distorted, not to mention childhood injuries, or running and jumping out of alignment. In fact, Joanna's son had infant torticollis, which means his head was stuck leaning to one side because of the way he was tucked up in the womb. He also had a jammed hip and severe growing pains off and on throughout his early years. Fascia therapy has helped him tremendously! I, myself was diagnosed with hip dysplasia. My hips ended up being the source of JRA pain so we have a lot to learn. Growing pains, I believe, happen when children outgrow their fascia. I've seen it time and time again and even in my own son, who grew a foot one year had horribly distorted fascia. We talked about this on our "Body Hacks" program on my Billionaire Healthcare® radio program. If you want more information on this topic, visit AshleyBlackGuru.com/The-CelluliteMyth for the link!

> Gotta say, it made me cry to see him running and goofing off so easily and pain free.
>
> -Elizabeth Snyder

As far as kids go, you can blast them if they let you, but it's far better to teach them to blast themselves. There's no age limit, just use common sense. With three grown kids of my own and two "adopted" kids, I feel like a professional mother. All my kids grew up around fascia treatment and FasciaBlasting family parties are a part of my life! **#AnotherBook**

We've had many reports from users about kids getting out of pain. Please follow their lead. Distract them with something they can do with their hands and reward them for sitting still. Start lightly. Teach them young. Passionate note: I believe this is part of being a good parent. Start them off right and teach them how to care for their bodies!

Elizabeth Snyder
November 10

Happy Update: My 10 year old son has Sever's Disease (a tendon issue) that causes him a lot of ankle and leg pain. It has prevented him from playing in organized sports and some days from playing outside. It broke my heart!! A month ago, I decided to Fasciablast him to see if it would help. I put him in a bubble bath and then lightly blast him twice a week. If you've ever tried to Fasciablast a kid, then you know it's like trying to Fasciablast a giggling Tasmanian Devil! He laughs his batoosh off!! He is SO much better! He can easily walk in the morning. His flexibility has improved and he can play dodge-ball as much as he wants. What a blessing! Gotta say, it made me cry to see him running and goofing off so easily and pain free. So that's what Ashley black does! She just wanders around making people HEALTHY & HAPPY!! XOXOXO

✦ ⬛ ★★★★★

CONTRAINDICATIONS, CONSIDERATIONS, AND RISKS

There are a million reasons to use the FasciaBlaster®. Here are the few reasons not to:

DO NOT BLAST:

1. If you are under the care of a physician and you are told not to. I cannot personally clear you for treatment so please read the following list and use your own judgement about when to start your program.

2. If you have a history of blood clots. THE FasciaBlaster® DOES NOT CAUSE BLOOD CLOTS. However, if you have any history of deep vein thrombosis or a blood clot, the clot could be held in place by dysfunctional fascia and if you open the fascia and release it, the consequences could be deadly. We do not play around with this. GET FULL CLEARANCE BY ULTRASOUND FROM THE DOCTOR.

3. If you are on ANTICOAGULANT blood thinners. It is not recommended for people who are on any type of blood thinners because of the risk of excessive bruising; however, if you talk to your doctor about fascia release and get approval, it's your choice with your physician.

4. In the presence of an acute infection or allergy symptoms. Rest and heal.

5. If you have uncontrolled congestive heart failure (CHF).

6. If you have Thrombophlebitis.

7. If you have Cancer or are currently in treatment.

8. If you have Neoplasms.

9. If you have abnormalities in your circulatory system or heart rhythms.

10. If you have Meningitis.

11. If you have open skin lesions.

12. If you have Pulmonary Embolism.

13. And never blast anything internally. Anywhere. At all. Ever.

CONSULT YOUR PRIMARY CARE PHYCISIAN OR SPECIALIST:

1. If you are pregnant.

2. If you are post-cancer.

3. If you've had recent abdominal surgery (less than three months ago).

4. If you have liver cirrhosis (because of the amount of detoxification).

5. If you have any abnormal abdominal pain. (Some swelling and tenderness is normal.)

USE PRECAUTION

1. During menstrual cycle (Listen to your body, don't over do it).

2. If you have a history of shingles. Shingles is a virus; viruses can lay dormant in the body. If shingles is trapped in the fascia, we theorize that releasing it could cause an outbreak; however, we do have people who have a history of shingles, including my mother, who uses the FasciaBlaster® regularly with no problem. Please talk to your doctor and make the decision yourself.

3. If you have a severe connective tissue problem such as fibromyalgia, Ehlers-Danlos Syndrome, or any issues that makes your skin sensitive, Use the FasciaBlaster®; however, there are very specific techniques that you can find on my YouTube® channel by searching the name of your diagnosis as the key word.

The rest of you, BlastOn™! HOWEVER, don't blast deeply or vigorously on the lymph nodes which are in the following areas:

- Arm pits (Axillary).

- Neck or jaw region (Supraclavicular or Cervical); be aware of location but you can continue to blast area.

- Groin (Inguinal)

 Kathy Ettinger
December 12

Heading to the bathroom to blast and my 11 year old says, don't go in circles or you will mess it all up! I said wow you do listen to me. Dad says well that's all you talk about . . . like Ashley Black for President and everybody's fascia will be healthy and healthcare will no longer be needed. I chuckled and said well that's a pretty good idea.

＋　　　　　　　　　　　　　　　　🖇　　　　　　　　　　✳✳✳✳✳

DIET

Everybody wants to know about the diet part of curing cellulite! When I was first talking to publishers about this book, many of them told me, "You must have a section on diet because that's what people want." The amazing publisher we chose was a little more open-minded, so here's your section on diet—there is no diet. For this book and for curing cellulite, cue the choir of angels because I want to make it abundantly clear, you do not need to diet and exercise to address cellulite. Having said that, diet and exercise *will* affect your body shape and size and your fascia at a cellular level, but that is for another book. Diet is simply not the core focus for successful elimination of cellulite. People who use diet to reduce cellulite are reducing the fat, which does reduce the appearance of cellulite. But it doesn't change the fascia. (It's like having *less* marshmallow crammed in the fence). I always recommend supporting the health of the fascia through clean eating, but clean eating alone won't break apart fascial adhesions. Here's how you can support healthy fascia through your diet . . .

FEED YO FASCIA

For you overachievers who are really looking for direction in your eating plan, here's a very high-level approach to nutrition and feeding your fascia.

Your body is made up of more than 37 trillion cells—blood cells, skin cells, muscle cells, and of course fascia cells. Not only is your body made up of cells, your body is made up of systems, which is how all these cells are organized to function within your body. Fascia is a system of the body that interacts with all systems of the body, so if you are doing something that is good for your endocrine system, it's good for your fascia system. If it's good for your blood, it's good for your fascia. We as a society need to stop picking everything apart so much and look at the big picture. Our entire body is a collection of cells and systems that are affected by the words of the day: "Clean eating." In a broad brushstroke, here is what clean eating is in my book: (And it is my book.)

1. No overeating. Your body can only absorb about 400 calories in a sitting. Don't overdo it and overtax your digestive system. When you overtax your digestive system, you overtax your fascia system. When you overtax your systems, you overtax your cells.

2. Eating 4 to 6 smaller meals throughout the day makes food easier to process than eating a couple of giant meals.

3. Eating clean also means eating real food, minimally processed, and in as natural a state as possible. If you want more specifics about eating in a way to reduce your size, please go online and read my "Get F'd" blog, which is the basis for a **#FutureBook** "The Diet and Exercise Myth." The blog addresses food, fascia, fitness, and FREEDOM! In the meantime, opening that fascia will help support any efforts you put into losing weight.

Now, there are specific things you can do to help support your fascia at a cellular level. On the next page is a chart of suggested nutrients to incorporate into your diet that support the restoration of fascia. This is not a comprehensive nutrition plan, but a high level brush stroke to show you the connection. There are lots of books out there on nutrition if you want to learn more. This book would be 8,000 pages if we went into detail on every single thing, so we didn't. Just take a look and do your own research on what interests you nutritionally.

NUTRIENTS THAT SUPPORT THE FASCIA

Vitamin	Function	Source
A	Promotes growth and repair of body tissues	Eggs, cheese, milk
Beta Carotene	Antioxidant	Carrots
D	Aids absorption of calcium, aids bone growth	Fish, dairy, sunshine
E	Antioxidant	Oils, peanut butter
C	Promotes healthy cell development	Broccoli, oranges, peppers, lettuce
Thiamine	Needed for normal functioning of nerves/muscles	Seeds, pork, beans, pecans
Minerals	**Function**	**Source**
Calcium	Bone growth, muscle contraction, nerve transmission	Dairy, fruits, yogurt
Phosphorus	Essential for DNA and cell structure. Works w/Calcium	Cheese, fish, spinach, leafy greens
Manganese	Necessary for normal development of connective tissues	Oats, clove, garbanzo beans, pumpkin seeds

DETOXING

Anyone who performs a dissection can clearly see that the fascia is the stickiest part of the human body. It literally feels as sticky as duct tape. It's not a stretch to believe that if you have something as sticky as duct tape all throughout your body that it can grab things and trap them—such as toxins, lactic acid, residual drugs from the past, food additives, air pollution, and so on. In fact, anything you've put in your body over the course of your life has the potential to be stored in your cells and released at a later time—even decades later.

This is another reason diet is so important. If you buy your food at a grocery store (which, unless you're Amish I'm sure you do) your "food" contains ingredients that your body has no idea what to do with. Have you ever read the ingredient list on the label of a frozen meal? Some will have more than 70 unpronounceable ingredients! Don't even get me started on fast "food." If you can't recognize an ingredient on a label, chances are your body can't recognize it either and what it can't recognize, it can store in your tissue. So, when you work on your fascia and begin to open and release the tissue, the toxins are being released as well. This is known as detox.

Toxins are all around us, in cleaning products, cosmetics, hair products, off-gassing of new furniture, carpets, flame retardants in beds, and so forth. Anything that comes from a factory is loaded with chemicals. In fact, there are more than 50 million chemicals registered with the American Chemical Society®—and counting! It's not all bad, but there's a lot of garbage out there being dumped into our bodies 24/7. It's no wonder that our skin can look like it's been crumpled in a trash compactor! I really want to make this point clear because people ask all the time, "How long is the detoxing?" Well, that depends on what you have in your body and what you are exposed to regularly.

When you blast, keep in mind that your skin is the largest detox organ of your body and you are stimulating it. It's also porous, so whatever is applied to the outside of your skin can seep in, and whatever is on the inside of the body can be pushed out through these tiny pores. This is why sweating and sauna are wonderful ways to detox through the skin.

When you blast, you are opening tissue and whatever was stored inside that tissue is going to be released. When toxins are released, they can present in a multitude of ways. Sometimes toxins are released immediately in the form of bumps or rashes. Other times they are filtered through the lymph fluid and blood. This gives your liver and kidneys a workout, which is another reason to heavy up on water. Other detox symptoms include: raised white spots, light-headedness, anger/emotion, changes in stool or urine color, soreness in the breasts and nipples, changes in menstrual cycles, spotting, swelling, strange-colored bruises, hot skin, flu-like symptoms, and in some extreme cases vomiting. And still, some people do not experience any symptoms at all. Search "detox" on the FasciaBlaster® private group page for personal stories.

This is not an all-inclusive list, and to be honest, the product is fairly new and every day someone experiences something new. Some can expect some emotional expression, being happy or sad, or even have flashbacks. I'll talk more about emotional detox in the next section. For now, just know that the body is so complex that you may have these symptoms. Please go check with your doctor for any issues that set off alarm bells. Above all, make sure you hydrate, and rest if you feel like you need to rest. Also, make sure you are in the right place in your life to take care of yourself while detoxing.

5 WAYS TO SUPPORT DETOX:

1. Drink plenty of quality water.
2. Supplement with Cellectrolytes™.
3. Support the liver (ie. milk thistle or castor oil packs).
4. Support the colon (ie. fiborous foods, probiotics, enzymes).
5. Sauna

We already talked about the importance of hydrating, which supports the kidneys. You may want to reread chapter 4. Also, many people report "elimination improvement" by simply blasting, but if you are having issues in this area, the first thing to do is increase water and magnesium. Magnesium is an electrolyte and supports more than 300 enzymatic functions. It also brings water into the stool to make it pass more easily and calms muscle cramps and spasms. There are several types of magnesium, but magnesium citrate specifically works in the gut and colon. Powdered forms or topical forms are often easiest to absorb and can help speed things along, if you know what I'm saying! My favorite is a product called Mag 07. Be warned, you may poop last year's corn. Ha!

Remember, if you are detoxing your body by opening your fascia tissue, the last thing you want to do is pour more crappy, processed, fake, devitalized food down your piehole. (FYI, "piehole" is Southern for "mouth.") The best way to support detox while Fasciablasting is to reduce toxic exposure by eating clean and making your environment as natural as possible.

Angela Riegel
August 2

I actually purchased the FasciaBlaster® of course because I saw the physical changes it could produce, but after reading and absorbing all the info and hearing about Ashley's story, I knew I needed this more for my health than anything. At the time when I purchased this. I had recently been diagnosed with hypothyroidism. I was sick and having a host of issues including high blood pressure and edema. Doctors could not find anything to help me since all of my blood work was within "normal" ranges. Since using the blaster my blood pressure is back to normal and I no longer have anxiety issues, hormone fluctuations are minimal, my hair loss has improved, and my circulation, oh glorious circulation is improving. I actually felt the energy rush from my heart pumping as it should have been. I have released so many toxins and trapped hormones that were wreaking havoc on my body. I have put my running on hold and been trying to focus on my structure and muscle isolations and have finally started seeing changes I can't explain in my body. I have a long ways to go, but the information and knowledge about my body, my diet, and exercise have been so valuable and have changed everything for me. This has done wonders for my self confidence and my closet is full of skirts and shorts that I have not worn on 20 years and that is completely priceless. I have even started to use this blaster on my husband who has restless leg and I can't wait to report what happens for him. Thank you Ashley Black for changing my life inside and out. I am becoming stronger in every way.

＋　　　　　　　　　　🗨　　　　　　　　　★ ★ ★ ★ ★

EMOTIONAL RELEASE

Emotional release is something that many people report, and it is not largely understood. As a therapist, I've actually experienced having visions of my clients' repressed memories and have had people literally sob when I started working on them. When I was running a Fasciology center this happened frequently and my staff members even reported getting sick after treating certain clients. Who knows if that was pathology release or emotional release, but we are complex beings, layered in an intricate spirit, soul, and body. There are studies being done on how the body, not just the brain, houses memory and emotion. You may have heard reports of people who have received heart or organ transplants having visions, dreams, and desires that the donor had. We can store memory in any cell of the body and the fascia seems to be one of those systems that can retain memories of all types. Just in my initial observations and data gathering, the front of the hips and stomach area seem to be the most emotionally charged. Not everyone experiences this, but if you do, it's normal and there is science to support it.

I am personally fascinated by this topic and anything that has to do with electrically charged particles or energy. When I'm alone and geeking out, these are the things I study and yes, I have a book in me about the subject. I'm not ready yet to introduce all of this to the world, but I'm super passionate about the topic and give my total love and support to all the women who are going through emotional release; whether good, bad, sad, angry, or revealing repressed memories. There's no real way to prepare for it. Just know if you feel "off" or emotional or touchy a few days after blasting, this could be what's going on. Take a deep breath, have some alone time, maybe journal a bit with a cup of tea. Take some time to pray and meditate about all the wonderful things you are doing for yourself and your body.

Nancy Schwartz
November 27

I bought the FB for the sole purpose of getting rid of my cellulite. I've suffered from cellulite for at least 5-7 years after having been a daily weight trainer since I was in my twenties & gaining weight over the years after slowing down in the gym and going through a couple very stressful life events. OMG!!! In just a few weeks, my cellulite is slowly going away. I've never been so excited to be black and blue because when the bruises heal, there's an incredible reveal!! I sent Ashley my pictures so she could assess the type of cellulite I had. She got right back to me and even suggested a routine to follow, and I got right on it! This isn't a race; it's a journey, and with consistency and faith, my body is absolutely transforming! I'm beyond excited!

Not only has this amazing tool started smoothing away my ugly dimples, but the way my entire body feels after each blast is almost unbeliev-able! I feel like I'm getting the best work-out I've ever had. While I'm blasting, I'm reflecting on my body image and how beautiful I am, no matter what condition my fascia has become. I blast with emotion and often cry because I'm doing something so personal and important for my health and well-being. With every challenging blast, sweat drops from my brow and afterwards I feel empowered and so accomplished. My entire physical body, from the inside out, feels better every single day! I feel like I'm detoxing away all the bad stuff that doesn't belong. My digestion is so much better and I'm never bloated like I used to be after a meal. My skin is glowing and my hair is shiny. My nails are stronger than they've ever been and I just FEEL on top of the world from blasting!

Honestly, at 47, now a single, empty-nester with grown children and a beautiful grandbaby, instead of feeling down for what I don't have around me every day and feeling less than attractive due to my weight gain, after having been so fit most of my life, I feel that this Fascia Blaster has literally given me a new perspective on life! My blood is flowing, I have energy, I feel like I'm getting my sexy back and I feel extremely empowered and educated on WHY my legs and arms and neck and belly look and feel the way they have for so long. But most importantly, I've learned and am still learning, from all of Ashley's videos and posts, that I don't need crazy diets, I don't need to buy a waist trainer, I need to exercise a certain way for healthy fascia, and I CAN and WILL have an incredibly healthy and beautiful, youthful body to showcase as a result of knowing how my fascia works and being able to use this amazing tool to keep it healthy, to sculpt it and to proudly walk around and LIVE as my very best and beautiful ME!

Thank you SO much, Ashley! My life is so worth living and looking and feeling on top of the world because of you and this incredible Fascia Blaster!

✛ ▰ ✶✶✶✶✶

FLUSHING

Flushing is a term used commonly in the sports therapy arena as a standard part of recovery and it should be a familiar part of your recovery too. After you blast, if you've had a really intense session, you want to either lightly blast or massage with your hands to move out the inflammation inherent with treatment. You can also flip the FasciaBlaster® over and massage lightly with the bar.

GRAVEL AND "NEW" LUMPS

We often hear, "I blasted and now I have a new lump." Well, actually, you don't have a new lump, you have a lump that has broken down into different lumps, or one that was buried under bound fascia that you couldn't reach or feel before. Remember, you have Cotton Candy stuck to an Ace® Bandage that is one big chunk, stuck to goop that is bound to your fat and muscles into one huge CLUMP. As you loosen it, you will feel the smaller chunks underneath. It's very common to go from tennis ball-sized knots, to marble-sized knots; then you break those up and it feels like gravel and then sand, and eventually, smooth. The FasciaBlaster® does not create new knots; it only reveals the ones that are already there. I call it **#BuriedTreasure**. Some women feel tight bands and lines. This is due to the different layers of fascia you are impacting and how you are breaking up the adhesions. Some adhesions feel like you are squishing a roach and others feel like duct tape being pulled apart. This is all normal! When you understand what's going on in your body, you will have confidence as you move through the process.

HORMONES

This is an issue related to a very complex science in and of itself. First of all, yes, you can blast while you are on hormone replacement therapy. Also, healthy fascia supports the endocrine system with or without hormone therapy. Blasting is excellent through menopause and helps keep the skin youthful by stimulating blood flow, nerve activity, and collagen production. We have women all the way into their 80s who are blasting!

We don't have a total explanation on how blasting impacts hormones but we want to make you aware that some women report spotting or period changes in the beginning. It does not continue to happen and we are looking into it. One theory is that estrogen is stored in fat cells and released in fat lysing. If you do experience a menstrual or hormonal change, know that it's temporary while the body is balancing out. If you have any questions or concerns about your hormones, please consult your doctor. We are diligently working on the research.

Christin Keen
June 22

I am 37 years old. I was born with rheumatoid arthritis and I have struggled with various autoimmune issues my entire life. Eight years ago, I gave birth to my daughter and my immune system started getting worse and worse. Three years ago, I had a full blown system malfunction, as I describe it. I do not have a better explanation. I had adrenal failure, thyroid and liver malfunction, my body stopped producing all my hormones except my estrogen which skyrocketed. My once athletic size 2, 120 lb body became a size 10 170lb body in 3 months time!

I was put on five months of bed rest with no change until this year when I discovered the FasciaBlaster® and Ashley Black. Through diet and blasting I have been able to feel human again. I am finally producing hormones, my hashimotos is almost in remission, my hypo is getting back in control and my adrenals are happy again. I am still working on lowering my estrogen and I cannot get a single pound off me no matter how hard I try, but the swelling and severe aching in my legs is gone. GONE! Having circulation back is an incredible feeling and I think I'm finally getting oxygen to my cells again!

On top of the smoother, healthier looking skin I have to say the most significant improvement I have seen in my life is how my daughter sees me. I never let her see me cry or struggle through the diseases and weight gain, I always encouraged that we love our bodies, even when I wasn't loving myself.

I still do not recognize myself in the mirror but now my daughter does! "OMG mom! You are starting to look like your old self again. You look beautiful! That blaster works!" She sees the change more than I do and she sees the new found confidence I feel and the depression and anxiety disappearing. She will never have to experience my struggle because I already have the answer. For that alone I owe Ashley Black and her amazing movement a million thanks! AND I'm only a month in! Can't wait to see the changes over the long term. My future is bright for me and my daughter thanks to Ashley Black.

+　　　　　　　　　　🖕　　　　　　　　　　✳✳✳✳✳

INJURY

There are two types of injury, acute and chronic. An acute injury, one that just happened, will leave you still at risk for a torn ligament or muscle, and you want to really proceed with caution. You would never directly blast on the injury but lightly blasting around it helps get blood in and out of the site of the injury, to speed healing. If you are really injured, such as with a slipped disk or sprained ankle, or are at risk of tears or fractures, please see your doctor.

For chronic injury, from repetitive motion like IT band syndrome or back pain, it is totally appropriate to blast all around the area, not directly on it. Start at the skin surface and see how you react. As you build up tolerance you can blast the old injury site and progressively get deeper, breaking up chronic knots and the like. It's important to note that the site of the pain is not always the site of the injury. For example, many people who have chronic low back pain have jammed hips, so the place to blast is in the hip flexors, the stomach and the front of the legs. Also, blast the line of fascia the injury sits on. (See the figures on pages 119–125.)

Injuries can be really tricky and it takes someone who understands Fasciology to help navigate recovery. But that's why you can send pictures and questions to me and my team!

SURGERY

We don't recommend blasting until 12 weeks post-op anywhere near the site or within 12 inches of the surgery site. You can blast the rest of your body. At 12 weeks use the acute injury guidelines and start by blasting around the area first. For both injury and surgery situations, you are going to naturally have more inflammation. Icing after blasting is even more important in these cases.

Sheila Helms
January 12

I purchased my FasciaBlaster® back in March 2015 after researching Ashley YouTube® videos I decided, "Why not try it? It's worth a shot!" I have always been into fitness but have had horrible cellulite no matter how small I got, or how hard I worked out. I would have never thought a small, one-time purchase would forever change my life!! I began to see changes I could have never achieved through diet and exercise alone. In July 2015, I was in a car accident. A man hit me going 70 miles an hour and I was in the hospital for 6 days with a broken back that now has rods and screws holding me together. As I had done before I contacted Ashley about using the FasciaBlaster® for recovery. The FasciaBlaster® was able to get me mobile faster, reduce pain, and overall got me back to a healthier than before the accident!

The FasciaBlaster® has helped with overall circulation, scar tissue, mobility, and has given me smoother skin. The FasciaBlaster® has changed not only my life, but my entire family! I have four kids and wonderful husband who uses the FasciaBlaster® as well. All of my kids play multiple sports so we use the FasciaBlaster® to help with muscle recovery, etc. I am an avid supporter of this product and tell everyone I know about the FasciaBlaster® and the benefits you receive, not just with cellulite but with over all health. Who knew one little tool could cure so many things!? Ashley is always so real and sincere when I have messaged her and she answers all my questions!! The FasciaBlaster® is a tool myself and family will always use. It is truly life changing!

MESH

Anyone who has had a repair with mesh anywhere near the surface would want to take their hand and brace the mesh, then blast around it. Do not blast directly over surface mesh. If the mesh is below the surface, like a bladder repair, you can blast lightly above the area. If you are not sure where your mesh is placed, consult your surgeon.

OVERCOMING PLATEAUS

First of all, let me say that women are crazy (myself included), and we all have a warped way that we see ourselves. You need to make sure you are actually plateauing and you really can't tell without doing the 5 Ps, particularly if you don't have photos. Also, if you are not using heat, you are fighting stiff tissue and won't make progress as quickly. You will achieve a level of success without heat but it will always be limited.

Now that you understand how the fascia is arranged in layers and you have learned about how it interacts with muscles, you can probably understand how a true plateau is possible. If you are going along and suddenly stop seeing results, that isn't necessarily crazy thinking. What you need to do is change your technique to break through to the next deeper level of fascia. Moving forward is about being consistent and digging deeper. The FasciaBlaster® claws have limitations, which is why we developed the Mini 2™ and the nugget. If you need to "get gangsta" on a certain spot, you need the end of the Mini 2™ or the nugget to dig it out. Again, we do provide professional coaching, and women who have gone before you can help you on our Facebook® page!

There's not anybody who can't get over a plateau as long as it's a true fascia issue. If someone is trying to flatten her tummy, but has fat behind the stomach, the FasciaBlaster® can't do anything about that because it can't affect what it can't reach. In this case, you need to diet and exercise.

Also, if you are taking two steps forward and two steps back by overtaxing your body, and doing high-impact activities, or wearing poor shoes that can reverse progress, you need to stop and think about your activities! Remember, bad biomechanics is the number one factor on bad fascia!

PTSD AND TRAUMA BLASTING

If you have been diagnosed with PTSD please know that a part of blasting is emotional release and flashbacks can occur. Please read the section on emotional release and work closely with your therapist to understand your triggers while blasting.

BLASTING WHILE ON MEDS

If you have questions about blasting and the meds you are on, please consult your physician. If you are on pain meds that inhibit a pain signal, please use caution so that you aren't blasting more intensely than you should since your pain signal is impeded.

SWELLING

Some swelling is inherent in the restoration of fascia because inflammation is a part of the healing response. Use the flushing technique or cold therapy to address inflammation. If you have concerns about excess swelling, please consult your doctor.

VEINS

The FasciaBlaster® is too new to formally make a claim about veins; however, we have had some pretty astounding "before" and "after" pictures, and we feel comfortable with the techniques we've developed.

To address varicose veins at the surface, blast lightly over the area and you can blast quite hard around it. When you first start to blast veins, they tend to look worse for a brief period but don't be alarmed. You are restoring blood flow, which makes the veins more pronounced. It's like the rush of water when you first turn on a pressure hose that calms down after a second. Even some people who don't have veins start to have what appears to be veins because the capillaries that didn't have blood flow before now do. With continued use of the FasciaBlaster®, we see them getting better and better.

Whether you have little veins come up in the beginning, or bigger veins you want to minimize, just keep in mind that the body has the ability to restore itself and you should be in a better position afterwards. BIG WARNING: Don't take a huge varicose vein and go ballistic because there is a chance you can burst a capillary, particularly if it's close to a bone. I have actually done this, although it did get better. Go slowly and gently. For spider veins, blast over gently and work your way into the deeper layers of tissue. I have a video on my YouTube® channel that demonstrates the techniques clearly, so please view it if you have any questions.

KEEPING YOUR NEURO IN CHECK

If you have a neurological issue, we have seen that stimulating neuro brings improvement. That's for a **#FutureBook**, but we are seeing it.

BLASTING SCARS

If you have scars on your hip, knee, C-section, ankle, shoulders—anywhere—blasting right over the scar releases that "puling" that can twist you out of alignment and it **#HurtsSoGood**! You can go in the direction of the scar, or cross-fiber (side to side), or use one little claw to dig out the painful parts. There is a video on my YouTube® channel that demonstrates how I break up the scar tissue on my hip replacement.

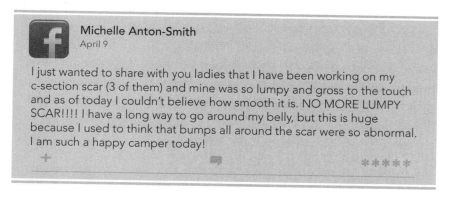

Michelle Anton-Smith
April 9

I just wanted to share with you ladies that I have been working on my c-section scar (3 of them) and mine was so lumpy and gross to the touch and as of today I couldn't believe how smooth it is. NO MORE LUMPY SCAR!!!! I have a long way to go around my belly, but this is huge because I used to think that bumps all around the scar were so abnormal. I am such a happy camper today!

HAIR GROWTH

Yes! There are studies that suggest that proper circulation and blood flow minimize hair loss or male patterned baldness. We've already said a #milliontimes that blood flow is the life source of the body. When those little hair follicles beneath the skin are nourished, they will do what they are supposed to do. We know the FasciaBlaster® stimulates proper circulation and improves blood flow, although direct results on hair regrowth have not yet been measured.

CANKLES AND CHICKEN LEGS

Cankles are nothing more than Beyond Bound™ jammed ankles. Blasting and opening the joints will improve cankles. Chicken legs are legs that are so restricted that the nerves are not firing properly and blood is not flowing properly to stimulate muscle growth. Blasting will help this too. In general, if you want something smaller, blast it. If you want something larger, blast and build it by weight training. **#FutureBook**

LIPEDEMA

We are in the early stages of exploring the impact of the FasciaBlaster® on lipedema. We have a small group on Facebook® with some very compelling "before" and "after" pictures. With improved circulation, blood flow, and nerve activity we are seeing results and hoping to study it more in the future.

ITCHING

It is very common to feel itching when blasting. The itching sensation is the nerves firing and blood flow. It's basically your body coming online and "waking up." Any time you feel this sensation, keep blasting through it! Also, wherever you feel a pain signal is a good place to blast. **#BlastOn**

FASCIA AND FLEXIBILITY

When working to improve range of motion, the fascia is as important, or more important, than muscle. Please refer to the Position test in the 5 Ps on page 118 to test range of motion and perform the stretches shown on a regular basis. In general, you always need to be heated before you stretch because if your fascia is tight and fighting you, then you are wasting your time. Always, heat, blast, then stretch. Stay tuned because there is much more to come on this topic! **#FutureBook #FutureDVD**

SCOLIOSIS

We have empirical data on this and many other conditions and diseases; however, it is really not for this book. Please visit the Facebook® FasciaBlaster® group to connect with women and hear their stories. **#FutureBook**

BUNIONS

We have anecdotal evidence that suggests blasting helps bunions. I am not going to say that using the FasciaBlaster® will prevent a surgery, but I hypothesize that it will at least help with symptom relief and improvement of function. As always, any information I give you is based on my experiences and advice. I encourage doctor's approval for flushing or myofascial treatment. Start out blasting lightly and progressing into treatment. The areas you need to concentrate on are the top and bottom of the foot, in between the individual foot bones, the arch, heel, and plantar fascia. Be sure to stretch the entire foot/ankle/calf musculature. This can be done through using a slant board, stretching the hamstrings, or doing the neural flossing technique which I have a video of. Go to AshleyBlackGuru. com/TheCelluliteMyth for the link.

FACE BLASTING

If you go on the FasciaBlaster® Facebook® page, you will see incredible "before" and "after" pictures of women who are FaceBlasting. At this point, the only thing I can recommend blasting is the jaw line and the forehead. The reason is because we are coming out with a product designed specifically for the rest of the face that is in the prototype and testing phase. WARNING: If you have clumped up fascia on your face and you blast it, you can create a dent or the appearance of a hole. Because the FasciaBlaster® lyses fat, you do not want to blast your cheekbones. Please be patient and work on the rest of your body; we will come up with appropriate products and protocols for the face. Don't worry, we'll help you slow or stop the aging process!

AUTOIMMUNE CONDITIONS

There is a direct link between autoimmune conditions and blasting. In fact, my whole life I've had an autoimmune condition, although I'm asymptomatic. Of course, this also needs a book of its own; however, for now we have recorded a radio program dedicated to this topic. Please see the Billionaire Healthcare® radio archive online for more information on this topic.

Susan Rosen
May 1

I am a 52 years young wife, mom, Breast Cancer survivor, who also has Hashimoto's Auto immune disease . . . And boy—have I struggled with cellulite! I bought the FasciaBlaster® because for most of my adult life I have been ashamed of my legs and butt. I have tried everything, diet exercise lipo wraps and leg make up. Ashley's system is working for me! I am seeing less ripples and dimples, my skin is smoothing out I am actually wearing shorts! Something I cringed at doing before.

More importantly, I have discovered that my Hashimoto's symptoms of dull dry flakey skin, muscle aches, and always having cold hands and feet have completely disappeared. The reason is because blasting has greatly improved my circulation!

Mentally and spiritually I have come to realize that I can change my health, my appearance, and my self-esteem. Ashley's videos, information, and the FasciaBlaster® community on Facebook® is changing my life! So thank you! The best is yet to be!

＋　　　　　　　　🗨　　　　　　　★★★★★

PRENATAL/POSTPARTUM

Yes, you can! In fact it's very good for your body to blast while pregnant and also when restoring your body. Please use similar precautions as in the "injury" section of this chapter and don't blast directly on baby. When postpartum, after a vaginal birth wait a minimum of six weeks to blast your abdominal area, and wait 12 weeks or *more* after a C-section. You can blast everywhere else, except the abdomen.

We don't know the implications or extent of estrogen release or detox when blasting so if you have concerns, wait until after breast feeding. However, it is perfectly safe for your body to blast all over. Much of this has been covered on the radio program Billionaire Healthcare® so please avail yourself of the archives online. (AshleyBlackGuru.com/TheCelluliteMyth).

Sarah Elizabeth Crawford
February 16

Wow! Where to begin . . .

I ordered my FasciaBlaster® through pure skepticism . . . I never dreamed I could get the results I have! I was 7 months post partum when I received it and got to work on right away. I had extreme hormonal weight gain, low energy from thyroid issues, an extreme overhang on my tummy and severe Diastasis Recti. Doctors told me I would never be able to lose the weight and after years of struggle, I started to believe them.

I noticed a massive change in my stomach in only 3-4 days! What I didn't expect was to feel a massive change in my energy, emotional wellbeing, healing of my stomach muscles, increased body strength and stamina, Not to mention the amount of inches I lost.

In one month of blasting, I lost 16 inches off my body. My Diastasis Recti has gone from 4 finger widths apart to 1 finger width apart. My tummy no longer looks 6 months pregnant! I have gone down several dress sizes. I have lost cellulite that I have had since junior high, my energy, libido and flexibility are off the charts . . . I feel A-MAZ-ING, inside and out!!!

I could never be more thankful than I am for Ashley and her protocols. I will truly never be the same!!! For a girl who had given up and just accepted a broken down body, I am moved to tears daily seeing myself in the mirror. I finally have "me" back . . . I am forever changed!!!

＋　　　　　　　　　　　　🖤　　　　　　　　　＊＊＊＊＊

PELVIC FLOOR

Very early on in my career, I worked as a consultant for the therapists at Texas Women's Hospital, and I've found there are a lot of misconceptions out there about the pelvic floor. I had no idea at the time just how many women had pain, vaginal pain, pain with sex, or some sort of vaginal dysfunction. Unfortunately, a lot of shame goes with it. But it doesn't have to. In my circle, we embrace women no matter what dysfunction they are trying to heal and we help them! **#BlasterSisters™**. It's time for women to come out find answers instead of hiding in pain and shame.

Having said that, strengthening the pelvic floor is more than just learning Kegels. It's about the neurological connection for all human movement 24 hours a day, seven days a week! You need to be working on your inner core and strengthening your pelvic floor, not just for sexual health, but for keeping your posture in check, and keeping your bladder strong so you don't have incontinence issues or need a mesh. Plus, there are a host of other benefits.

First of all, you need blood flow and nerve activity to the area, which not only will bring healing and restoration, it will also increase sensitivity. Focus on blasting the inner thighs and around the pubic area. Also, the fat pouch right between the belly button and the pubic bone is made up of fat just like any other area of fat on the body and can be blasted to reduce puffiness. If you have loose skin, gently blast externally but DO NOT blast on the inside. In short, blasting improves blood flow which improves libido and makes you youthful and tight! **#GoBeaverBlasters**

LOOSE SKIN

While the FasciaBlaster® can do some amazing things, it is not a complete miracle worker. (Depending on how you define miracle.) We know that the FasciaBlaster® does tighten skin by restoring blood flow from the inside out. The blood carries the very proteins that nourish the skin, and support collagen production, and if the blood is not getting through the fascia your skin is not going to be as healthy. By restoring the fascia, you are supporting your skin cells and you will see a difference.

Some people experience droopy skin because of blood flow, some people have it because the muscle fibers are being partially choked out by tight fascia causing them to atrophy. Atrophied muscles look like fat. There is no shape or firmness to it. If you are someone who has had muscle fiber choked out by fascia but your skin appeared tight because of the tight fascia, you might experience a temporary appearance of loose skin but at the end of the day, you need to activate the atrophied muscle.

To activate atrophied muscle:

Think about the muscle and contract or squeeze the muscle without doing any movement. If you cannot isolate the muscle this way, you are not ready to move or add weight to it.

Activate through movement. If there is any question about proper movement, please go to AshleyBlackGuru.com/TheCelluliteMyth. (You know where to go by now.)

Keep in mind, skin alone cannot be tight; you have to muscle in place. For example, if you have a flat butt and loose thighs, you need to build muscle to hoist the buttocks up. If you have loose skin from having a baby, and you've had scar tissue that has been held down by fascial adhesions, your skin may be hanging over. First you would open the fascia to release the tissue, then build the muscle to tighten the skin as much as possible. If you are Beyond Bound™ or Gummy Bear, you will have loose skin after opening the fascia, so be prepared to activate those muscles and flush the area.

Chelsea Stillman
September 29

Since I started using the FasciaBlaster® even just with Ashley's tutorials I've experienced changes I didn't anticipate. I originally bought it for cellulite reduction and to break down fat cells, which did happen over the course of 30-60 days. But after the first use, and yes a nasty bruise, my hip was completely free of pain after a 13 car pile up that knocked me around in the passenger seat. I was a mess and with the FasciaBlaster® my jammed hip finally popped in place on it's own after my chiropractor spent 3 years unsuccessfully trying to adjust it. This accident, along with a stressful year of school and another stressful year of workplace abuse triggered fibromyalgia. I was in debilitating pain, and honestly, I was fighting suicidal thoughts a few times. I had to change jobs which helped me to get my sanity back and then one of my close friends told me about the FasciaBlaster®. As I continued to use it, my overall pain and stiffness dissipated! It no longer felt like my skin was going to rip apart with every movement. Now, instead of laying in bed in the morning because I can't move, I lay in bed simply because I want to. That freedom of choice is priceless! I also noticed, as did one other client I treated, that abdominal blasting has improved menstrual cycles. We both experienced easier monthly symptoms and less cramping. Some of you may have seen the photo of my treatment for "turkey neck" on the group page. When my client saw how it firmed up another client's abs, she asked if it would help her neck. I'll tell you right now, I didn't wait to ask Ashley if it would work. I tried it on myself first. When I saw a difference in my own neck I then did the same thing to her and to my surprise it tightened up on the first try! This tool has also saved my career in massage therapy because even with the job change my pain had me thinking about switching careers all together and honestly I don't know what else I would do. Even now as I heal from reconstructive surgery on my wrist the FasciaBlaster® is keeping my circulation healthy since I'm not able to do much. My weight loss

> journey is a long one but not as long as it would be without Ashley and the FasciaBlaster®!
>
> +   💬   ★★★★★

WEIGHT GAIN

First of all, throw away the scale. Second of all, don't argue. It's not that I don't want you to measure yourself or track your progress, I just want you to use an accurate, consistent, reliable measure. The number on the scale doesn't accurately tell you anything about what's going on inside your body. I am 100 percent anti-scale. There are just too many variables, including how full your colon is. Not only that, a person can be losing fat, building muscle, hydrating better, increasing blood volume, drop four sizes, and never lose a pound on the scale, yet it's the most common measure of progress.

Not only that, when you are blasting, you are going to have some inherent inflammation as part of the healing response as the fascia tissue is being remodeled. For a day, or two or three, you can look slightly larger but as the swelling goes down, you will be tighter and smaller over the longer term.

The other type of gain is when you start to activate atrophied muscle. You are pumping the area full of blood so that the muscle fibers can grow. When you reestablish blood flow and neurological connection it will make the muscle toned and toned muscle weighs more than untoned muscle, but it's not necessarily going to increase in size just yet.

When your body is working more efficiently, your metabolism increases. You are restoring your body, and it will help with fat loss and getting in shape if that is what you desire. However, if you are experiencing dramatic weight gain or loss, it is outside what the stimulation of the FasciaBlaster® produces and you should go see your doctor. If you are having a seriously negative effect, start by checking your food or hormones.

So, the next question is: How should you measure your progress?

1. Take pictures!
2. Measure yourself with a tape measure.
3. Check progress via the 5 Ps.

FASCIA RECOIL

If you've had a restriction in your joint, which means it's basically been glued together by fascia, when you open it up, it doesn't mean the joint is going to be stable; in fact, it's quite the opposite. Opening the fascia means you have access to the joint and can build muscle. So some people feel great using the FasciaBlaster® and then overuse it. More is not better! A joint that's just been opened is like a joint that just had surgery. If you overdo it, you can cause a spasm, throbbing, or severe swelling and redness. If you have extremely tight tissue, open it up a little at a time and do stability exercises in between, particularly in the neck, back, hamstrings, and knees. These parts of the body often get "glued" down because they are major players in stability and movement. When the blood rushes in, you experience temporary swelling, but if you go slowly and strengthen the joint in between blasting sessions, you will have a better, less painful result.

VERY IMPORTANT: If you have an area of the body that's been bound and restricted for a long time, be sure to blast all around it to make sure trapped fluid doesn't get stuck and inflamed. For example, there was a lady with neck pain who decided to vigorously blast her neck. Well, the neck is about six inches long and that's the only area she did. She did not blast the full lines of fascia or her head, arms, trapezius muscle, and so on. Her entire neck swelled and she couldn't move her head for three days, because the healing inflammation had nowhere to go. She temporarily felt like she was injured, but when she started blasting around the area, she saw results.

BE SMART! Don't attack an old injury or scar tissue all at once. Do a little at a time and activate the muscle in between.

MUSCLE FATIGUE

Keep in mind that you are going to feel a lot of different things when you blast. You might feel tingling, which is better blood flow and nerve signal. You might go to the gym and find your muscles fatigue faster. That's because you've opened muscle fiber that hasn't been worked in a long time. The tight fascia was covering a muscle weakness. This happened to me when I first invented the FasciaBlaster®. I had a Fasciology center at the time and I was training my staff to use it on me. I got about two hours of blasting in one day, mainly on my legs, which feels amazing!

Then I went to the gym and did the routine I had been doing for literally year: leg extensions, curls, abduction, adduction, and calf raises. I was in muscle maintenance and did the exact same weight I had done for almost a year. After my workout, I could hardly walk and I was intensely sore for seven days. The reason was that the blasting had opened muscle fibers that had been asleep for years and I didn't even know it! Those muscles came online and my "regular" routine wore them out! If you get super sore or feel weaker after blasting, it's because you have more access to your muscles and more muscle fibers are actually firing. My suggestion is to open a little and strengthen a little, and go through the process slowly.

THIS IS ONLY THE BEGINNING

As you may have already gathered, the rabbit hole is massive. There are least ten or more books that can be written based just on what we know now. While fascia has been studied for generations, never before in history has there been mass fascia treatments in order to gather empirical data for more in-depth studies. In less than a year, the knowledge of fascia has spread like wildfire and people are hungry to learn more. Now that we have the FasciaBlaster®, we can pull data and appropriate sample sizes very easily. Anyone interested in a research project anywhere in the world, be it for school, science, or a foundation, we would be interested in exploring the effects of the FasciaBlaster® on everything.

 Michelle Barris
November 2

I wanted to start by thanking Ashley for this opportunity to have our voices heard, and to share our personal stories. I wanted to also let Ashley know how much gratitude, appreciation and respect I have for her and what she is doing for us.

Empowerment, Knowledge, Transformation and Community, these are the words that summarize the journey that I did not know I was about to embark on when I purchased the Fasciablaster. Like many, I don't have a million bucks, but I want to look and feel like a million bucks and have tried almost everything to try to come close to this goal.

Empowerment- I am a full-time, working, mother of five beautiful children. I am active and health conscience. I have tried basically every diet and exercise program that is out there, but never really happy with the final results. I have struggled with "cellulite" and "Asian Booty Disease" for several years but have not been able to make lasting changes. I was intrigued and excited about the possibility that I too could have a smooth heart-shaped butt! I didn't realize that I was going to be empowered with the tools, knowledge, support to make changes in my life, my body, my health and my family.

Knowledge- So no one ever told me cellulite doesn't exist. No one ever told me that the very diet and exercise things I was doing were actually counterproductive to what I wanted to achieve. Not only has this journey been full of knowledge about how to improve myself aesthetically, but there is so much science behind it (yes I am an undercover nerd!) This knowledge has made me change the way I feel about my body. I have a lifetime of struggle with body hate of this or that, but learning the science behind this from Ashley has turned this into appreciation and self-love.

Transformation-Yes, I was vain and bought this originally for cellulite, but little did I know it was about to transform my whole body, save me from shoulder surgery and help my family. My body was breaking down- foot, hip, knee, shoulder, elbow (I now know why and how these were all connected!) I recently found out that I have calcification of my rotator cuff tendon and suffered from lack of range of motion, weakness, constant pain, etc. So 6 months, 2 procedures and months of PT later, they suggested surgery. I did not want to go down the long painful road of surgical recovery, so I recruited help to blast my shoulder. I was amazed that after just one session, I was able to move my shoulder in ways that I was unable to- even with months of PT. I have used it on my daughter (who dances competitively) to help with her pain, my husband who suffers from a low back injury and my dad who has arthritis and constant pain. They are all amazed at how they feel after I blast them, but I am not- I know the power of this little stick.

I know this is long, but I wanted to make sure I said something about the community. Typically when you buy a product, you are left to your own to figure it out, trouble shoot, etc. However, the community that Ashley nurtures comes with support, love, encouragement, helps keep you going. They are there to answer your questions, share your successes, and motivate you. Yes, I did not realize that my life would be changed when I bought this product. I didn't realize that I could spend hours talking to other women about fascia and they would be just as interested in what I had to say. I didn't realize how addicting blasting would be or that I could after all of these years be at peace with my body and love an appreciate it for the amazing things it does! Thank you Ashley Black from the bottom of my heart!

✛ 💬 ✶✶✶✶✶

10

From Heaven, to Hollywood, to Households

Here's how it happened . . .

*"Obstacles cannot crush
me. Every obstacle yields to
stern resolve.
He who is fixed to a star does
not change his mind."*

–Leonardo da Vinci

LOSING CONTROL

It is an ordinary night in a small town in Alabama where a 6-year-old girl, joyful, energetic, and filled with curiosity about the world, prepares to go to bed. She has shiny blonde pig tails, hazel eyes, and a smile that lights up the room. Even at such a young age, she is a highly intense and competitive athlete. After school, when other children are going home to watch TV, this young lady is in the gym practicing gymnastics. By age 5, she'd already mastered back handsprings on the beam and was well on her way to realizing her dream of becoming an Olympian and following in the footsteps of her favorite athlete, Mary Lou Retton. After nearly four hours of practice, her mother would have to practically drag her out of the gym—but she would always negotiate one last vault or one more spin around the bars before leaving.

"Say good night to Daddy," her mother tells her, snapping her out of her daydream about her practice session earlier that day. The satin ruffle on her pale pink Holly Hobbie® nightgown kisses the floor as she bounds across the room into her Daddy's arms. A quick squeeze around the neck and a butterfly kiss and she is off to bed. It was a night like any other night, but it would not stay that way for long.

As the hours pass, this sweet, innocent child lay sleeping peacefully in her bed while visions of gold medals danced about in her head. The house is dark and quiet except for the ticking of the family clock down the hall when suddenly, she is violently awakened. Her eyes pop open and fear grips her heart as an intense burning pain begins to flood her entire being. It is as though she is being mercilessly attacked with a fire iron and all she can do is lay there overwhelmed, and frozen in the agony.

The torture continues to intensify as she somehow manages to roll out of her bed and onto the floor, unable to scream or even speak. With tears streaming down her bright red face, she drags her stinging body along the carpet until she reaches her parents' room down the hall.

This was my earliest recollection and introduction to losing control of my own body and my own health. That day, I was introduced to the cruelest of perpetrators. One from which there was no hiding. Although it wouldn't be realized until years later, my diagnosis was Juvenile Rheumatoid Arthritis (JRA) and mine was as acute and aggressive as it comes.

Growing up, the pain would come in waves and I never knew when it was going to hit, or how long it was going to last. Imagine a perpetrator coming for you, torturing you, stealing from you, and attempting to destroy you. He leaves and you know he's coming back again and again. You don't know when, or where, and you have no protection. It was like living in a horror film with no way to turn the channel, no hero, and no escape. But, it never stole my drive or passion for gymnastics. What it did do was force me to find another way to become better. I had to work even harder to stay on top of my game, and I was determined to push the boundaries of my body even further. I never accepted the limits my diagnosis tried to put on me.

At some point in life, we've all felt limited. We've all felt out of control. We've all experienced overwhelming pain and the emotions that ensue. This is my journey, the foundation for my life's work. It is filled with intense highs and indescribable lows. It was this rollercoaster with my body that forced me out of the box of mainstream medicine and fitness, both traditional and alternative, and drove me to find another way to make my body do what I wanted it to do. What I didn't realize at the time was that my discoveries would not only make me better, faster, and stronger, but they would make anyone better, faster, and stronger. Over the years, I've watched countless individuals from every walk of life—executives, grandmothers, pro athletes, and celebrities—with virtually every type of body goal break the boundaries of what they thought was possible and morph into a better version of themselves.

As you read this story, know that your best is better than you think. You do have control over your body and appearance. You can become better. You can look better, you can feel better, you can have better muscle access and better movement. You can approach your body in a whole new way, and with your new understanding you can reach new levels of fitness and performance and change your appearance beyond your wildest dreams. It is my great honor to share with you the knowledge I've learned through my personal struggles and victories.

RISING FROM THE ASHES

As a child, I lived in a constant state of fear and panic because I could never predict what my body would do and I had absolutely no control over it. The doctors managed my disease with anti-inflammatory medications and hip aspirations, which was a procedure where they inserted a long needle into my hip capsule to drain out the excess fluid.

Some doctors encouraged me to remain as active as possible; others suggested replacing my hips and restricting my activities. I chose the former because the latter was a certain path straight to a wheelchair, which they predicted would be my fate by the time I was 30 anyway. Besides that, a hip replacement would only last 10 years, which would mean multiple surgeries over my life, and I wasn't even a teenager yet.

I spent my grade school years on and off crutches as the JRA would run its cycles. The other kids teased me relentlessly. They called me a faker because I had no visible signs of anything that they could comprehend, piling more cruelty on to an already brutally ruthless disease. They wondered how I could be on crutches one day, and then winning a regional gymnastics meet, or running stadium stairs the next. I firmly believed that if I stayed active and continued to participate in sports that I might not need the crutches so much, and I could somehow beat this thing at its own game. The more it tried to take me out, the more driven and active I became. I did gymnastics, swimming, baton twirling, and anything that would keep me moving. I was also careful to follow a rigorous stretch program for my sports, already understanding this was a key to my pain management. My love affair with human movement had begun, perhaps because I appreciated the ability to move far more than those who had never faced the possibility of not moving.

Even as early as the fifth grade I was looking for a way around the medical system. I didn't like someone else determining my fate for me. I was sure there was another answer than what I had been given. I can remember being in the doctor's office one day and they wanted to take more blood for more testing. I didn't want another needle stuck into me so I bargained with the doctor and said, "If I can do back flips all the way down the hall, will that prove that I'm well enough and don't need further testing?" Sure enough, I did back flips down the hall and avoided the needle that day.

In retrospect, I was already discovering how to manipulate my fascia. I would track my food, which was not something done in the late 70's. I would massage myself in heat and take ice baths. I read in magazines about the programs sports stars were doing and I tried them all, having no idea that my destiny had been set into motion.

The foundational principles that would be the basis for my life's work were born out of sheer desperation to feel like me—athletic, strong, and agile—when this disease fought so hard to take me down. I needed to take control of my body, my life, and my shaky, scary world. I was driven with intensity head-on into my true calling and purpose in this world.

In high school, I absorbed everything I could about the human body and how to make it stronger. I was determined to fill my battle arsenal with ammunition to fight my predator, lurking in the shadows. By 16, I taught aerobics at a local health club and became a personal trainer. I was fascinated by body building, and not only was I a "gym rat," I was a gymnastics rat, and a cheerleading rat, and an all-things-athletic rat. My entire life revolved around making my body stronger. While the hip aspirations continued to be a routine and necessary part of my life, the movement and activity somehow eased the attacks, keeping their frequency down and the severity at bay. I continued gymnastics through college and started competing in dance. It was all part of my pain relief protocol and it was working. By this time, my arthritis was almost un-detectable. I had gotten myself in such amazing physical condition that even I almost forgot I had it.

As a young adult, I got involved with a nutrition company both personally and professionally. I began to understand how nutrition, hydration, hormones, and stress affected my muscles and joints at a cellular level. Now diet was also an integral part of my new health routine. I had turned myself into a human guinea pig, experimenting on myself out of necessity. I was by all accounts "healthy," but the vicious perpetrator was merely hiding and in weak confinement. I had put him in chains through movement, meds, and aspirations, but he was laying in wait with one last elegant, cruel trick in his bag. I was about to face my worst nightmare that would have made the attacks of earlier days my pleasure.

THE NIGHTMARE BEGINS

In my mid-20s I was living in Houston, Texas, married and blessed beyond imagination with two amazing children, a toddler and a newborn. It was just after the birth of my second child that my life's course was permanently altered. For starters, even though I was in the best shape of my life and taught a kickboxing class the day before I went into labor, the birthing process triggered an onset of pain unlike any I had experienced. The worst was in my right hip. The doctors assumed that the hormones in my body from the birth were causing the flare-up since arthritis is an autoimmune disease. I hadn't ever sought alternative care before this time, but one of my nurses was the equivalent of a chiropractor in another country. She would do some simple reflexology and touch my legs with a healing hand. This brought a level of relief I wasn't getting from the anti-inflammatories and pain meds I was taking.

When they released me to go home I could barely care for my newborn daughter. I struggled just to lift her and certainly couldn't carry her around the house. A simple feeding left me utterly exhausted beyond comprehension. The pain in my right hip was completely unbearable. I tried to hide my agony as much as I could, not wanting to be a burden to my family and hoping that it would somehow go away. When I finally could take it no more, my husband took me to the emergency room, where I received yet another routine hip aspiration—or so I thought.

I had undergone this ordinarily uneventful procedure countless times before, but this time would prove to be tragically different. At first, I began to feel better and was released to go home, but later that evening, I awoke in a nightmare. My entire body was paralyzed and it felt like fire ants were crawling through every single nerve ending. Every time I even attempted to move, it felt like I was literally being struck by lightning or sitting in the electric chair. I didn't know what was happening, but I thought for sure I was dying. My husband rushed me back to the ER and I was immediately admitted to the ICU.

Not surprisingly, I had a fever, which elevated to an alarming 108 degrees Fahrenheit as I lay there silently screaming in utter agony. I was spiraling into medically uncharted territory for my disease, and I knew my life was slipping away. The pain was winning, I was completely lost in it, and the worst was yet to come. I was so overwhelmed on every level that my brain couldn't even process it. All my tests were inconclusive; the doctors had no idea what was wrong with me. I begged for an epidural, which was denied because it would have killed me. The doctors were baffled and couldn't fathom that I was in as much pain as I was reporting, so they sent three psychiatrists into my room to evaluate my mental condition. I remember literally grabbing one of them by the coat and shouting, "Look! I have no history of mental illness. My family has no history of mental illness. If I'm telling you my body feels like it's on fire, IT'S BECAUSE MY BODY FEELS LIKE IT'S ON FIRE!'"

I was beyond desperate and against my natural grain, I pleaded and demanded that they cut me open and figure out what was going on. I literally felt like I wouldn't last more than a couple of hours. When the medical team surgically opened up my hip, they stepped back in shock at what they saw and the stench of death that filled the room. Even though the bone scan showed no infection, what they found was a septic mess. A voracious infection had spread throughout my marrow system and had eaten two-thirds of my hip. My diagnosis was osteomyelitis, a bone-eating bacteria—a staph infection. Only about one-third of the people who contract osteomyelitis escape with their life.

The doctors could never explain why I had such intense pain all over my body, but I now know it was because the bacteria had spread to my spinal column. As I explained in chapter 2, the spinal column is encased in a special straw-shaped fascia where all the nerve roots protrude. The pain was an intense fascia signal, an "alarm", coming from all over my body. At the time I had never heard of fascia, but I was experiencing its mightiness first hand. I didn't know that the key to my pain would become the key to my purpose. This "event" would lead from a full blown fascia mutiny to my becoming a student in understanding its function and a pioneer in techniques to restore it.

When they discovered the bacteria, I was blasted with the most powerful antibiotics known. My body was ravaged not only by my merciless disease, a lethal infection, and now by a massive dosage of drugs pumped directly into my heart—drugs used to ease the suffering of the dying. My Dad blessedly slept at my bedside, hitting the morphine drip every ten minutes, pushing aside a busy architectural practice and foregoing his own sleep for weeks, while my Mom was raising my newborn daughter and toddler son.

While in a drug-induced coma-like state, I prayed to die and be released from my pain. I begged them to let me die on a daily basis. To me, the world outside of my pain no longer existed. I experienced heart failure twice. I was only in my 20's yet I had two near-death experiences with all their glory. No choir sang, no majestic symphony played, but yes—it all flashes before your eyes and the light is there. That, however, is another story and a whole book unto itself.

My stay in the hospital was pure torture. On top of the relentless pain, I was heavily medicated and couldn't even sleep because they were constantly running tests or drawing blood. My body actually became stiff, like rigor mortis was setting in. I couldn't shower. I couldn't go to the bathroom. My stomach looked like I was nine months pregnant because they couldn't get a catheter in, and I was covered in oozing bed sores because they couldn't move me. I wondered if I would ever go home.

Eventually, they did get the bacteria under control, and I begged them to release me. I literally felt like I was going insane. I looked like a crack addict—pale and skinny, weighing about 30-40 pounds less than the healthy body I had when I was admitted. When I held my arm up, the skin and muscle draped over the bone like a piece of fabric. The physique I had spent my entire life developing was nothing more than Jell-O® mush.

On the last day I was in the hospital, I was totally overwhelmed. As I was wheeled out to my car, the sunshine hit my face for the first time in what felt like an eternity. I recall thinking that this must be how soldiers feel coming home from war, struggling with the transition from such a horrific world back to normalcy. But for me, things were hardly going to be normal. My dance with death changed me. I had morphed back into a mother and now a scientist, a spiritual being, and a fierce warrior. I had survived this brutal attack, and this time I was back with a vengeance. This thing, this disease that tried to take me out was about to see the Ashley Black "A" game, and now the real battle had begun.

I went home with a morphine drip in my leg, an antibiotics central line to my heart, and a walker. I also had about a four-inch leg discrepancy. My legs were the same length but my hip was jammed into the socket and my socket was jammed into my ribs. Remember, two-thirds of my hip was gone. My entire right side looked like a derailed train, and I moved around like I was 90 years old.

Even though the bacterial infection was under control, I would still need another six months of intense medication—medication that is no longer legal. As I entered my home for the first time, friends and family were gathered and I was greeted by two small strangers—my children. My baby girl was a couple of months old and she didn't even know me. My 2-year-old son was scared of me and all the tubes hanging from my body. I must have been in a state of shock because I remember that just having extra bodies in the house nearly gave me a panic attack. I was certainly in no mood to party.

My sister coated my hair in baby oil and wrapped it in a shower cap for two days just to detangle the mess of the hospital stay. I received training at home from nurses about how to run my own IV at the kitchen table. Twice I did it wrong and instead of meds going into my heart my blood was being drained out of it. I was a complete mess and, as you can imagine, caring for a 2-year-old and an infant while in this condition was next to impossible. My husband was at work all day, and I felt very alone not knowing how I was going to manage all of this, but somehow, I did. That same determination that drove me to compete at the highest level of athletics drove me to compete for my life again. I even had the courage to have baby #3 only 21 months after this unbelievable event. There was never a moment that I accepted my condition. I knew that somehow, some way, by the grace of God, I was going to get back to normal and get back to being me.

THE PATH TO HEALING, THE PATH TO DISCOVERY

As the months passed, for the most part I was healing except for the open bedsores on my back from being immobile for so long. My mother somehow got the phone number of the nurse who performed therapy on me in the hospital after my daughter was born. She would come to my house to treat me, and she was the first person to ease the vise grip this pain had on me. Exercise was out of the question. I could barely roll onto my side. The doctors had prescribed physical therapy, but the therapists would come to my house and even they didn't know how to help me. I was too debilitated to do their most simple moves. I needed pre-physical therapy therapy. Really, what I needed was a "Yoda®" to take control of my case and show me the way. But there was no Yoda®. I was discovering that I had fallen into a black hole in the medical community. I felt alone in my head, like I was floating on a board in the middle of the ocean with no land in sight. How do you strengthen your legs if you can't stand up? I knew I had to figure out another way to do it. So, in true Ashley Black form, I developed my own version of *Sit and Be Fit*™ or really, "lay and be fit." After all, conquering pain and pushing the boundaries of my body is what I had done my whole life.

I knew I had to start by simply getting my brain to talk to my body again. So, I would lay in bed and do a series of isometric muscle contractions one at a time, and it was exhausting. Without movement or resistance my muscles would burn like I had just done an intense "leg day" in the gym. This was the beginning of my full-blown, sound-the-fire-alarm, self-prescribed rehabilitation program. For months, my life revolved around strength training. Eventually, I was able to lift weights in bed, and do some of the exercises I had done before. I practiced my gait with my walker until I could walk on my own. At the time I saw this as a victory, but the leg discrepancy eventually caused a lumbar scoliosis that I would need to reverse later on as well.

My program kicked into high gear when I met a local chiropractor. He lived in my neighborhood and when he heard about my situation he offered to help. He was the first person to aggressively manipulate my soft tissue after this insane process. He basically started with a glorified massage the first visit, but I made more progress that day than I had in months. His hands calmed muscles that had recoiled into spasm, and this was the catalyst for my journey into alternative medicine.

I finally began to take my workout out of the bed, out of the house, down the street, and to the gym. I got to the point where I could start working at the health club again. I needed to do something to get out of the house, and I remember actually teaching a class with the antibiotic cords still hanging out of my chest.

At this time, Pilates was virtually unknown, but I heard about an instructor course for this "new" type of exercise. I heard that many of the exercises could be performed lying down, which is exactly what I needed because standing up was still so incredibly dysfunctional for me. I spoke with my manager about becoming an instructor and the club began flying me to San Francisco on the weekends to learn from one of the most seasoned Pilates practitioners in the country. This is when I began to incorporate nontraditional exercises into my own therapy as well. Within two months, I became the national trainer for the entire chain of health clubs across the country. Back at my location in the suburbs, I began modifying the moves, improving upon them, and developing my own hybrid-Pilates. Because the results from my version were far more astounding than traditional Pilates, my classes were packed. They actually had to enlarge the studio to meet the demand. It wasn't long before my location was doing more in revenue from Pilates than every other location in the entire country combined.

I also began flying to Phoenix to receive flexibility therapy. My goal was to even out my legs so I could walk without a limp. In my first stretch session, the practitioners had my legs filleted opened like a frog being dissected, and for the first time I almost felt normal. As they were pulling and tugging my body back to something that resembled human, I remember tears streaming down my face, without sadness, as a wave of emotion uncontrollably overtook my body.

I traveled some more and tried a whole host of healing methods, settling into a simple program of self-treatment of soft tissue, fascial and neurological stretching, neurologically based exercises, and total body integration. These methods fall outside the scope of both traditional and nontraditional medicine and are the framework for a new genre I call Fasciology.

My self-treatment was working. I began feeling better. I was functional, and I was on the upswing to becoming athletic Ashley again. The health club continued to fly me to San Francisco to take Pilates courses, and I was refining my understanding of human movement. The last piece of the puzzle for me was figuring out how to stabilize my spine and pelvis, in hopes that the results of all this therapy would last. So far, nothing I had done was the magic key to the castle. I ferociously read anatomy text books, body work journals, medical journals, and anything I could get my hands on that would shed some light on how to find this last piece of the puzzle.

One day, while attending the Pilates training, I had an epiphany. The seasoned instructors, who had been doing Pilates for years, were demon- strating a movement in the upper level curriculum. This one particular perfect-bodied instructor, whose movement was so elegant and strong, performed every single movement flawlessly—except one. She had to ask for someone to spot her so she could perform this certain movement. I remember watching the movement and thinking that it required a combi- nation of all things athletic: strength, flexibility, agility, balance, breath, and form. I couldn't understand how this "goddess" who had performed every other move with perfection needed a spot for this move. To me, she was the holy grail of biomechanics! My competitive spirit came out, and I vowed to conquer that particular move on my own completely, without a spot. As it turned out, this would prove to be the birth of my exercise regime that I now teach to my clients and followers.

When I returned home from the training, I began to study and analyze this move. I wanted to know why it was different and what made it so difficult. The answer I now know was simple. This move required stability of the *inner core,* which was not forced to fire in any of the other movements. Like a mad scientist, I began reconfiguring the Pilates machine, shortening and lengthening the levers, altering the drag co-efficient, changing the angle of gravity, and breaking down the biomechanics one step at a time. I analyzed that move spinal segment by spinal segment—what was stabilizing and what was moving—and simplified it piece by piece. I cracked out the anatomy books and figured out "prep moves" for performing this elite movement. I didn't know it at the time, but what I had discovered was how to intentionally activate muscles that the medical community thinks fire unconsciously. All I knew was that I was getting stronger and within three weeks I was able to perform "the move." More amazingly, for the first time in more than a year, I was able to walk without a limp.

The "prep moves" I developed are a series of body reprogramming training techniques. I began to refine this series of movements and created about 15 different moves that seemed to be the most effective. I went back and showed my chiropractor what I was doing, and he was intrigued enough to allow me to use his back-pain patients and his Olympic® track athletes as my guinea pigs. Amazingly, the back patients began to experience less pain, and the Olympians were getting stronger and faster and jumping higher and longer. I knew I was on to something revolutionary in body training. It was a game-changer!

A MARKED MOMENT

Not long after my exercise discovery, a weekend came that would forever change the course of my life. At the time, I just thought I had these really cool exercises and I would make a business out of it and maybe teach some classes. But I was about to learn how earth-shattering my discoveries were.

I was invited to present my newly discovered movement techniques to a women's track team at a major university, whose coach had the most winning record in NCAA® history in women's track and field. I was attending a course with brilliant minds and other inventors who were well-established in the world of sports training and medicine. As each "Yoda®" presented his or her therapy, technique, or training, the coaches, staff, and I would watch in awe.

I couldn't wait for my turn. At the time of this conference, I had recently had my third C-section and was only about two months postpartum. I had followed my own system throughout the pregnancy, so I was ready to demonstrate the moves when it was my turn to present. I began to demonstrate my program, holding the positions, and showing variations of how to apply them in sports. I was able to hold the proper positions and even gesture with one hand as I spoke to the crowd, never once getting tired. Everyone was very impressed and anxious to try my exercises.

The first person who tried to get into the base position couldn't even hold it for a few seconds. He was one of the coaches and rationalized that he was more out of shape than he thought. So they brought up an athlete, and surprisingly, he couldn't hold the move either. They brought up athlete after athlete, including two Olympians, who could not hold the position, much less perform sport-specific variations. How could this be? How could a woman who was two months postpartum from a C-section—whose abs were sliced in half—how could this woman out-train an Olympic® track athlete? I looked around that room and it was a very ominous moment. We all realized the potential, that these top athletes had to become even better with my routine. It was at that moment that I realized how remarkable my discovery was. It was the first time that I really understood that this was my calling and I had to bring this knowledge to the world.

THE BEGINNING

I spent the next year testing my exercises in different positions on people from all walks of life. I trained most of these people inside that tiny chiropractic clinic. As I tried to teach the movement techniques, I quickly discovered that many people had structural limitations that prevented them from performing the exercises correctly. For example, their spines wouldn't rotate, their hips were too tight, their ankles were jammed, their knees misaligned or their muscles were just completely bound and tight. These were perfectly "healthy" individuals, with no complaints to speak of, yet their movement was limited. At the end of the day it was quality of their fascia and the placement of the bones collectively that caused restrictions they weren't even aware of. It is restrictions like these, if not attended to, that over time snowball into compensation after compensation that first limits ability and then becomes injury. I knew my next step was to gain the knowledge and skill to correct and align the physical human structure.

I set off once again on an educational journey, traveling all over the country. I studied different philosophies of manual tissue work and internal stretching through courses created by MDs, chiropractors, physical therapists, massage therapists, healers, and personal trainers. This was a huge step for me in terms of truly being able to make a difference in people's lives. I took a blend of techniques from all of the coursework, I brainstormed with the geniuses who surrounded me, and I used trial and error daily to refine the processes, yet there were still pieces of the puzzle missing and in a big way. So many of the actual techniques didn't make sense and didn't seem to solve the problem. So, instead of turning to other people's methods, I studied the anatomy and physiology myself and came up with my own! (Shocker, I know.) I studied dissection from the "fascia" people and realized that the body should be dissected layer by layer versus by pulling pieces out like the childhood game Operation®. I began to understand the fascial lines explained in chapter 6. I started to feel and understand the 4 types of fascia and the puzzle pieces were beginning to fit easily. The epiphany: We missed it. The science of fascia had been lost, and I was on a mission to revive and evolve it.

At this point, I had many overwhelming success stories: Fasciology in its infancy, and a hot fire in my belly driving me to understand how to put it all together in a science-based manner and deliver it to the world. First stop: research. Getting grants was next to impossible, but I had wonderful clients who experienced life-changing results who generously donated their personal money to have fascia studied. I conducted a sports performance research project with a well-known physician in Houston. We studied the effect of my methods on healthy, middle-aged golfers, and we found the results to be undeniable and measurable. The players averaged an astounding 20-yard gain on their drives in just six short weeks!

Since this discovery, we've see the same consistent, measurable results with athletes across the board—a track athlete surpassing a personal best, a field goal kicker with more consistency, a pitcher with increased velocity, a marathoner with a shorter time, and the list goes on.

We also measured the voltage in surface muscle output with my techniques and compared it to the other types of core exercises like crunches and Pilates. The result: with my methods, the muscles fired more effectively and more consistently than any other exercise. All the while, clients were benefiting in other ways and the testimonies were piling in. Clients who came in with shoulder pain ended up shaving time off their marathon. Clients whose professional athletic careers were about to end because of long-term repetitive injuries ended up with the MVP award the next year. The success of what I was doing was attracting big names and major attention. My heart would race any time my cell phone showed the caller ID of an unknown area code because I knew there was another NFL®, NBA®, or PGA® athlete calling, or another high-end businessman who was going to fly in to get my "secret sauce." These professional athletes and executives were initially coming in because of injuries, but soon discovered with my recommendations all of their movements were better, their sports performance was better, and overall, their lives were better. Every day was like Christmas—a new client, with a new problem, in a new field. It was a new challenge every day and Fasciology was the answer!

In the early days, I would think to myself, "How in the world is this former sickly gymnast, recently back from the grave, doing what I am doing? How in the world can I be the freaking expert? How on Earth could a world-class athlete who can afford to go anywhere in the world end up in MY 12x12 office?" Did I really know the secret? As it turns out—I sure did! Not only that, but I was now personally in the best shape of my life. I was lean and mean, doing what I loved, and happy to wake up every single morning.

I was later asked to lecture at the MLB® Sports Medicine conference and I absolutely wanted in! At that conference, I met so many doctors and trainers who were living their dream. One particular doctor out of Nashville took a special interest in what I was presenting. About a week after the conference, he called me and said that he wanted to fly a Pro-Bowl® NFL® star in to see me. In fact, he wanted to bring several players with him. It suddenly dawned on me that I needed my own Fasciology center. I had been so immersed in what I was doing and the quality of the work that it never even occurred to me that I might need some more space.

So in 1999, I built my own center. The mission: to help as many people as possible. I hired physical therapists and seasoned trainers. I shared an office with an orthopedic surgeon, a nutritionist, and chiropractor. Each employee came from a slightly different background and brought something new to our understanding of Fasciology. For years, my center was a virtual sharing of knowledge, a petri dish of discovery for the how, what, when, where, and why of the body. I was a magnet drawing the creative minds to the think tank.

One of the greatest advantages of the center was that our clients came from all walks of life. They ranged from teenagers to people in their 90s. We saw brain stem injuries, multiple sclerosis, chronic pelvic pain, tendonitis, and the list goes on and on and on. Amazingly, as different as their symptoms were, the method to solving these "medical mysteries" was always the same: Fasciology.

I used to think that in order to expose the world to Fasciology I would need to have countless centers across the globe. But something interesting evolved that changed my perspective and approach. We had so many people coming in from out of town, or people who just had limited time or access to the center, that I was constantly being asked, "What can I do on my own?" This forced me to explore a self-treatment component. When given the proper knowledge and tools, clients following an at-home method were able to obtain similar results to those who came to see us in person. This demand was the catalyst for the design of the FasciaBlaster®. It was shortly after I had prototypes in the hands of my clients that we really began to see the power of this tool. Almost every single user of the first set of prototypes was a pro athlete. In fact, the very first FasciaBlaster® user was Travis Hafner, who played baseball for the Texas Rangers®, Cleveland Indians®, and New York Yankees®. But as each of my players saw results, their team members each wanted one, too. And then their friends and family members and it just started rippling out and demand started to quickly rise. I told in chapter 5 how the girlfriend of one of my players reported that her cellulite was gone, and that was when my whole life's trajectory changed. With the onset of social media and the Internet, I had the vehicle and the gas, so to speak, to take this knowledge around the globe. It was time to bust out of the walls of the center. Everything was right in front of me and all I had to do was figure out how to harness the power and develop a plan.

At the same time that all of this was taking off, I was pursuing a publishing deal on another book about fascia called *Becoming Superhuman*. This is when Joanna came into my life through a mutual friend. I'll let her tell you that story in her own words (instead of her telling it in mine)...

A Tale of Two Authors

It seems everybody wants to write a book. However, not everyone can. Sure, they may know what they want to say but they just don't know how to "say it" on paper. That's where I come in. As a writer and ghostwriter, people approach me all the time, asking how to write a book or if I'll help them. So, when a mutual friend first told me about Ashley, her "friend who was trying to write a book," I had an internal eye roll. Most people really aren't that serious, especially if they don't already have the money and following to pull it off. If they do have the money and following to pull it off, they have a book agent. However, I loved and respected this friend enough to roll on over to Ashley's center to learn more about this "body stuff" that Ashley does.

The day of our scheduled meeting, I arrived at her Fasciology center address in this low-key shopping center right off the highway. It has a small drug store on one side and a cheap, fast food chain restaurant on the other side. There were several high-end luxury vehicles and a black Suburban® parked out front. The double doors to her suite were understated with dark tint, and I thought, "What in the world does this woman do and what have I gotten myself into?"

As I walked in the door, I entered a large, open space that somewhat resembled a night club. It had dark walls and ceiling, red leather lounge seating in the waiting area, and photos of celebrities posing with this cute blonde girl everywhere. There was also a life-sized skeleton and anatomy illustrations hanging on the walls. Eight or ten massage tables were lined up in two rows where people were having Fasciology sessions right out in the open.

I sat down to wait for Ashley and awkwardly looked away from the action. Either I was running early or Ashley was running late, but when she bounded in the door in her tank top and cargo pants, she brought a ball of sunshine with her. She saw me and came straight over, a giant smile beaming from her face, and she gave me an even bigger hug. "Hi! I'm Ashley! I'm so excited you're here! Come over to my table and let's get you taken care of." I had no idea what she was talking about or what I even needed, but I'm fairly adventurous and curious and she seemed sweet enough. "Stand up straight," she said. I stood there before her and she kind of shook her head. "That's straight to you?" She told me about all the places where I had structural deficiencies that related to my 20 years of knee pain. In the same manner one would say, "Pass the salt" she said, "I'll have you taken care of in about an hour." Now, I had done everything known to man to get rid of my knee pain: surgery, prescriptions, therapy, training, chiropractic, and so on. I couldn't believe what Ashley was saying but sure enough, she delivered. After about an hour of a not-so-pleasant Fasciology session my structural deficiencies were drastically improved, and I got up and did a lunge for the first time in my life without pain. I exploded into tears and was completely hooked on fascia—and Ashley—in that moment. I knew in my heart that we were Divinely connected to bring this life-changing information to the world.

The first thing we did together was write a book proposal. Typically this takes a couple of months and we did it in 2 weeks flat. Our book agent was completely hooked and we were quickly talking to top publishers. Even my contacts at the publisher loved the book and felt the power of what I was doing. We went before their finance board three or four times, but when all was said and done, they declined the project. Even though I was well known in sports circles, they said, "Ashley, you're not famous, and no one has heard of fascia. So you either need to get famous so you can present fascia, or fascia needs to get famous and we present you, or both." So I said to myself, "Okay. If that's what it takes." At least I knew the assignment—go make me famous or go make fascia famous. Done deal. I knew it wasn't going to happen in Houston. As much as I love my city, it's just not where people become famous. As far as fascia is concerned, the work and research is primarily being done in other countries. So when I looked at all my options, I decided on L.A. because I thought I could get the answers that I needed there.

I was mostly looking for the right connections and thought maybe I would meet someone who could help me do an infomercial, or someone would tweet® about me, or a celebrity would partner with me, or I could do a reality show, or *something*. There wasn't a "Famous" pill or a *"Becoming Famous for" Dummies®* book or I would have been all over both. (Of course, I could write that book now!) I just started knocking on every door I could think of and, well, I encountered a lot of closed doors. First of all, an infomercial took $600,000 to make plus millions to launch it. Nope. The TV shopping channels didn't want my product because bruising could result. Yep. I even tried *Shark Tank®* and they just weren't interested in what I had to offer. Of course, now they call me all . . . the . . . time. I sat there thinking, "I'm sitting on an atomic bomb, and I can't get an airplane to go launch it." I was back to the "how to become famous" drawing board; meanwhile, my client list was growing more interesting by the day.

I wonder if those doors closed on me because I just wasn't ready yet at the time. While I had all these ideas about how to "get famous," which I now know is just a way of assuming a position of influence so I could present this science to the world, I wanted to do it the right way. I wasn't really that comfortable with the notion of becoming famous, in and of itself, because I've seen how people accost celebrities in public and scrutinize their every move. Your job becomes your life and you have to always live "on." I'm a mom, and a huge surfer, and the last thing I wanted to do was give up my privacy or time for the things I love. What I quickly realized is that in order for me to fund my own research, join the ranks of billion-dollar businesses, and launch a product globally, the fame and the money are necessary. I've always lived on a budget and being a single mom with three kids, this was all a huge step out of my comfort zone, but I had to do it. So the big question in my mind was, how do you find this careful balance between presenting your ideas on a mass scale and becoming famous, yet not being in it for the money or fame?

Well, as it happens, that's exactly what I learned. I went to Hollywood to find a connection to help me launch my brand, and then I found out that connection was me! I was already equipped to do what needed to be done; I just had to make the link within myself.

I had this impressive clientele list in L.A. and as I worked on their bodies, I asked a lot of questions. As a Fasciologist, I don't have to verbally coach movement like a trainer. I don't have to be quiet and create a "spa" feel. When I'm working with someone and opening the fascia, it's not really relaxing and frankly, it doesn't feel so good, so people enjoy conversation as a distraction. I talk to my clients about everything! I've had hours and hours of one-on-one conversation with real billionaires (**#WithaB**) who are completely below the radar. I've talked candidly with the most well-known celebrities and royals of our time and with people who used to be famous who learned how to manage it all and keep their private lives private. I worked with people like me who are "edu-tainers" and talked about how they got their starts. They were all more than happy to share their stories and secrets with me. ('Course, when you have your elbow in someone's back, they will tell you anything you want to hear!)I learned personally from a vast group of highly successful people, and what I learned was how to be successful by first accepting my place in the world, and then embracing the truth that I had the power to make a massive impact. In fact, it was my ordained duty to leave the world a more empowered, educated, and better place.

What I needed was more people to experience what I had experienced and feel compelled the way I do to help others and pay it forward. It all happened in a way I never imagined. After establishing a pretty significant following on Facebook®, I decided to start a private, women-only coaching group where women could submit photos and ask questions and I would personally help them. What happened was a blend of the perfect timing, perfect setting, perfect platform, and a message that was desperately needed to be delivered. The women who joined were craving information, then putting it into practice, and motivating and helping other women in a way I've never seen before in my life. They were taking this science into their own hands and transforming their worlds. They were learning, practicing, and sharing. They were whipping out that FasciaBlaster® on their husbands, family members, and loved ones and taking it to educate their doctors! It's been like a wild fire, unstoppable and beyond any prayer I could have ever prayed. I didn't need an info-mercial. I didn't need a celebrity partner. I didn't need a reality show. All I needed was other women like me—and I found them everywhere. In fact, this extraordinary group of early adapters are so powerful and passionate that they drove THIS book to #1 on Amazon® in just four hours. This is "fame" redefined and the fascia movement, as it has turned out, is as powerful as a title wave. Fame and fascia . . . check, check!

I'm so overwhelmed by the support of the Blaster Sisters™! We are truly in this together. I'm not the perfect-bodied product model. I'm just like the women in my group working hard to stay in the best condition I possibly can. With my past diagnosis and my crazy travel schedule, blasting has to be my number one priority. It keeps me functioning. I can be completely pain free and fine one day, and if I have a couple of drinks, a cramped flight, then walk 3 blocks in heels and eat pizza crust; I can go from totally healthy to blinding pain with migraines and barely able to get out of bed. Blasting and the knowledge of fascia brings me back. Not only that, spreading the word and educating people involves a lot of unfamiliar beds and not having access to the foods I need. I can't always stay on top of it. Some weeks I'm on point doing hot yoga, sauna, and blasting daily. Sometimes a week goes by, or two or three, and I barely have time to address my scar. We're all in the same boat trying to find balance. For me, it's about having control to get back on track when I can. One thing is for sure, I don't ever allow myself the headspace to freak out over cellulite or pain, because I can fix it. That's the control and peace of mind I endeavor daily to instill in my followers.

Every day I am blown away by the testimonies, photos, thanks, praise, and sincere appreciation. People send me gifts and ask me all the time what they can do for me and the reality is, the best gift that anyone can give me is to live fully in what you've learned and to share this knowledge with anyone who needs it. **#AshleysArmy** has been assembled and we are going on our march. Grab your FasciaBlaster® and everyone you know because together, we can inspire accountability and change both personally and socially. From the bottom of my heart, thank you for being a part of this amazing movement. We are changing the world through . . .

. . . knowledge, empowerment, inspiration.™

BLAST ON™!

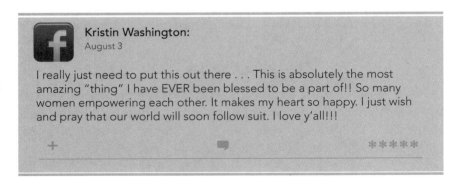

Kristin Washington:
August 3

I really just need to put this out there . . . This is absolutely the most amazing "thing" I have EVER been blessed to be a part of!! So many women empowering each other. It makes my heart so happy. I just wish and pray that our world will soon follow suit. I love y'all!!!

Bibliography

"The Causes and Treatments of Cellulite," WebMD® Medical Reference Reviewed by Stephanie S. Gardner, MD on May 31, 2015, http://www.webmd.com/beauty/cellulite-causes-and-treatments#1

Ignaz Semmelweis, from Wikipedia™, the free encyclopedia, Internet Resource Information, https://en.wikipedia.org/wiki/Ignaz_Semmelweis

The American Experience, The Rockefellers, **John D. Rockefeller, Sr.**, [This biography of John D. Rockefeller Sr. was written by Keith Poole, Professor of Political Science at the University of Georgia, as part of a course on Entrepreneurs and American Economic Growth.] http://www.pbs.org/wgbh/americanexperience/features/biography/rockefellers-john/

Walt Disney®, Immensely Successful High School Dropouts, http://under30ceo.com/11-immensely-successful-high-school-dropouts/

Cirque du Soleil®, History, https://www.cirquedusoleil.com/de/news/history

Fascial Dissection Courses, Tom Myers, Structural Fascia "Trains & Stations," Learn more: https://www.anatomytrains.com/fascial-dissection-courses/

Treatment for Plantar Fasciitis, WebMD®, Plantar Fasciitis Treatment, reference from Healthwise, © 1995-2015 Healthwise, Incorporated. Healthwise, Healthwise for every health decision, and the Healthwise logo are trademarks of Healthwise, Incorporated. http://www.webmd.com/a-to-z-guides/tc/plantar-fasciitis-treatment-overview#1

Leonardo da Vinci, Leonardo da Vinci: anatomist, Roger Jones, Editor, The British Journal of General Practice, Copyright © British Journal of General Practice 2012, https://www.ncbi.nlm.nih.gov/pmc/articles/PMC3361109/

Ancient Rome Bathing, last modified on 26 November 2016, From Wikipedia™, the free encyclopedia, https://en.wikipedia.org/wiki/Ancient_Roman_bathing

There are 37.s Trillion Cells in Your Body, Rose Eveleth, Smithsonian.com, October 24, 2013, http://www.smithsonianmag.com/smart-news/there-are-372-trillion-cells-in-your-body-4941473/

American Chemical Society®, 50 Millionth Unique Chemical Substance Recorded in CAS REGISTRY, September 8, 2009; https://www.acs.org/content/acs/en/pressroom/newsreleases/2009/september/50-millionth-unique-chemical-substance-recorded-in-cas-registry.html

What makes vessels grow with exercise training?, reference from American Physiological Society, Journal of Applied Physiology, Prior, Barry M., Yang, H.T., Terjung, Ronald L., September 2004, vol. 97, no. 3, http://jap.physiology.org/content/97/3/1119

RESOURCES

The Cellulite Myth and the Fascia Movement are supported by a dynamic, online community with countless videos and resources. Everything you need to know is available at: AshleyBlackGuru.com/TheCelluliteMyth

Index

THIRD-PARTY TRADEMARKS AND THEIR RESPECTIVE OWNERS